Magnetism in Topological Insulators

Vladimir Litvinov

Magnetism in Topological Insulators

Springer

Vladimir Litvinov
Sierra Nevada Corporation
Irvine, CA, USA

ISBN 978-3-030-12055-9 ISBN 978-3-030-12053-5 (eBook)
https://doi.org/10.1007/978-3-030-12053-5

Library of Congress Control Number: 2019931905

This Springer imprint is published by the registered company Springer Nature Switzerland AG
The registered company address is: Gewerbestrasse 11, 6330 Cham, Switzerland

To Daniel and Anastasia

Preface

From the time that semiconductor spintronics was recognized as a mainstream in solid-state physics, the search for new materials that allow the manipulation of magnetism by an electrical field has become widely discussed. Films made of topological insulators (TI) are relatively new material systems in which the voltage-controlled electron spin polarization at the surface can be used in device applications either by itself or as a tool that excites magnetic switching and oscillations in an adjacent ferromagnetic layer.

The topological insulator discourse calls on the concepts of topology, differential geometry, and other complex areas of contemporary mathematics. This book is aimed at studying magnetism in TIs on the level that fills the gap between postgraduate courses typical for students in physics and electrical engineering. The prerequisite for understanding this material is a general undergraduate-level knowledge of quantum mechanics and solid-state physics.

The book deals with the relation between the topological structure of Bloch functions and the properties of TIs in electric and magnetic fields. Common properties of TIs are studied based on a particular TI example of $Bi_2(Se, Te)_3$. In Chap. 1, we study the energy spectrum of surface Dirac fermions in thin films. Chapter 2 is devoted to the Berry and Zak phases in quantum mechanics and in solid state in relation to various Hall effects. Chapters 3 and 4 discuss ferromagnetic proximity effects and the topological magnetoelectric effect. In Chap. 5, special attention is paid to the s-d interaction between surface fermions and magnetic impurity. The magnetic indirect exchange interaction between a pair of magnetic atoms in a 3D TI is considered in Chap. 7. In Chap. 8, we briefly discuss the current and prospective applications of TIs in semiconductor electronics and solid-state magnetic devices.

Irvine, CA, USA V. Litvinov

Preface

Contents

Chapter 1
Energy Bands in Topological Insulators

Topological insulators (TI) are identified by their gapless surface electronic states with the energy lying inside a bandgap of the bulk band structure. This makes the crystal conducting on the surface and insulating in the bulk. To study the magnetic and electrical properties of TI materials and to understand what features make them different from conventional semiconductors, we start by studying the energy spectrum of an example material system $(Bi, Sb)_2(Te, Se)_3$. Bismuth chalcogenides is not the only class of materials that shows a topological response to external electric and magnetic fields: we could also mention other TI materials such as II–VI compounds $(Cd, Hg)Te$ and also IV–VI semiconductor alloys $(Pb, Sn)(Te, Se)$ which are called crystalline topological insulators. Semiconductors from II–VI and IV–VI families are TI examples that are less convenient for experimental studies due to the small energy gaps and also because of the strongly non-stoichiometric growth that makes it difficult to pin the Fermi energy inside the energy gap of bulk crystal, where the topological effects become observable.

The electrical properties of Bismuth chalcogenides and their device applications in thermoelectricity have been intensively studied and thus all the details of their energy spectrum, electron, and phonon scattering mechanisms have long been known. What makes them special as TI solids? We start from the microscopic approach, analyzing the band structure of bulk Bi_2Se_3 and the specific features of the surface electron states which differentiate them from conventional solids. A phenomenological approach that relates TI general properties to band topology and symmetry will be discussed in the next chapters.

1.1 Spin–Orbit Interaction in Bismuth Chalcogenides

The bismuth chalcogenides family crystallizes in a rhombohedral structure with point group D_{3d} and space group $R\bar{3}m$. The structure consists of the set of quintuple layers illustrated in Fig. 1.1. Quintuple layers are coupled by the weak Van der

V. Litvinov, *Magnetism in Topological Insulators*,
https://doi.org/10.1007/978-3-030-12053-5_1

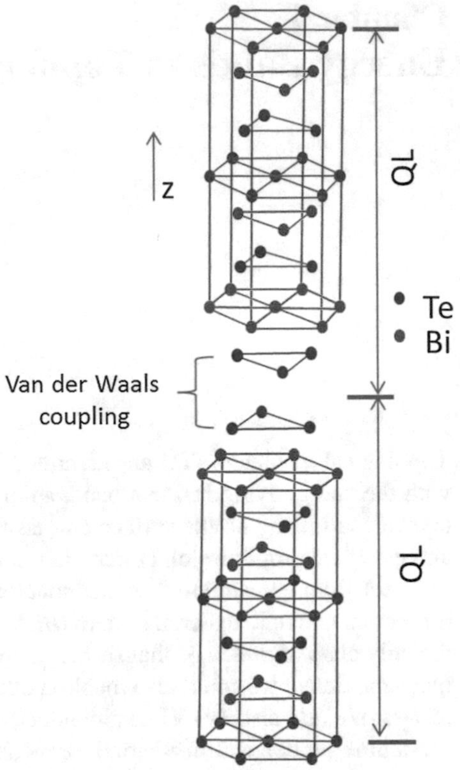

Fig. 1.1 Bismuth chalcogenide crystal structure. QL stands for a quintuple layer

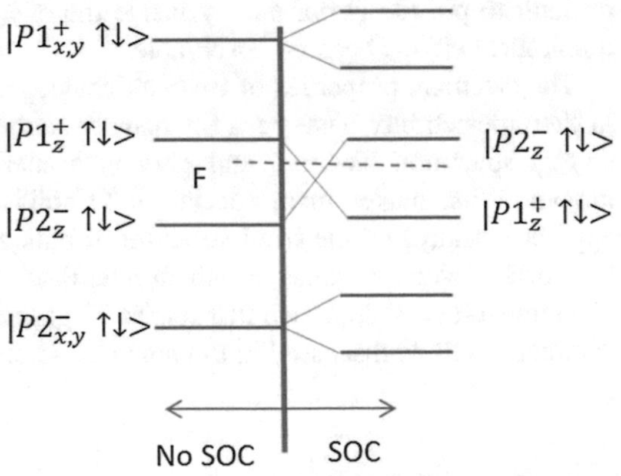

Fig. 1.2 Spin–orbit coupling inverted energy levels corresponding to hybridized *Bi* ($|P1\rangle$) and *Se* (*Te*) ($|P2\rangle$) p-orbitals

Waals interaction that results in easy cleavage along planes {0001} perpendicular to axis *z*.

Relevant electron energy bands have energy close to the Fermi level, they stem from Bi- and Se- p-orbitals and have been calculated in [1, 2]. The origin of conduction and valence energy levels are shown in the left panel of Fig. 1.2, where wave functions $|P1\uparrow\downarrow\rangle$ and $|P2\uparrow\downarrow\rangle$ stand for double spin degenerate linear combinations of the atomic Bi- and Te- p-orbitals, respectively. Wave functions have parity (±) with respect to inversion symmetry and also account for the energy splitting between $P_{x,\,y}$ and P_z levels caused by the strongly anisotropic crystal field. Spin–orbit coupling (SOC) in the lattice potential does not cause the spin-splitting as far as both the time reversal and the inversion symmetries hold (right panel in Fig. 1.2). Conduction and valence band edges exchange their positions if the SOC is large enough. That is the band inversion induced by the strong SOC that creates the non-trivial band topology and turns semiconductors Bi_2Te_3, Bi_2Se_3, and Sb_2Te_3 into TI.

1.2 Electron Spectrum and Band Inversion

In solids, the Schrodinger equation for Bloch functions $\Psi_{nk} = u_{nk}\exp(-i\boldsymbol{kr})$ can be reduced to the equation for Bloch amplitudes as follows:

$$H_{\boldsymbol{k}} u_{n\boldsymbol{k}} = E_{n\boldsymbol{k}} u_{n\boldsymbol{k}},$$

$$H_{\boldsymbol{k}} = H_0 + \frac{\hbar^2 k^2}{2m_0} + \frac{\hbar^2 \boldsymbol{kp}}{m_0} + H_1 + H_2,$$

$$H = \frac{p^2}{2m_0} + V(\boldsymbol{r}),\ H_1 = \frac{\hbar}{4m_0^2c^2}\boldsymbol{\sigma}\cdot\nabla V(\boldsymbol{r})\times\boldsymbol{p},\ H_2 = \frac{\hbar}{4m_0^2c^2}\boldsymbol{p}\cdot\boldsymbol{\sigma}\times\nabla V(\boldsymbol{r}), \quad (1.1)$$

where $\sigma_{x,\,y,\,z}$ are the Pauli matrices, e, m_0 are the free electron charge and mass, respectively, $\boldsymbol{p} = -i\nabla$ is the momentum operator, and $V(r)$ is the electron potential energy in the periodic crystal field. The fourth and fifth terms in the Hamiltonian (1.1) represent spin–orbit interaction. In order to find the eigenvalues E_{nk}, one has to choose the full set of known orthogonal functions that create the initial basis on which we can expand the unknown Bloch amplitudes u_{nk}. In the vicinity of the Γ-point, the set of band edge amplitudes u_{n0} normalized on the unit cell volume Ω can serve as the basis wave functions (Luttinger-Kohn representation). Within *kp*-perturbation theory the third, fourth, and fifth terms in the Hamiltonian (1.1) are treated as a perturbation.

Conduction and valence bands stem from two double spin degenerate levels which present the band edges in the vicinity of the Γ-point in the Brillouin zone (Fig. 1.2, right panel), so the basis functions $u_{n\mathbf{0}}(\boldsymbol{r})$ can be taken as $|P1_z^+\uparrow,\downarrow\rangle$ and $|P2_z^-\uparrow,\downarrow\rangle$. The matrix Hamiltonian in u_{n0} - representation is written below:

$$H_{nn'}(\boldsymbol{k}) = \int \langle u_{n0}(\boldsymbol{r})|H_{\boldsymbol{k}}|u_{n'0}(\boldsymbol{r})\rangle d\Omega, \quad (1.2)$$

where the integration is over the volume of the unit cell.

Energy corrections from remote bands enter in diagonal elements of $H_{nn'}(\boldsymbol{k})$. They are proportional to k_i^2 and contribute to effective masses turning initial levels into energy bands. Taking the basis functions in the sequence, $|1\rangle = |P1_z^+ \uparrow\rangle$, $|2\rangle = |P1_z^- \uparrow\rangle$, $|3\rangle = |P1_z^+ \downarrow\rangle$, $|4\rangle = |P1_z^- \downarrow\rangle$, one obtains the Hamiltonian matrix with k_i^2-accuracy:

$$H_{\mathbf{k}} = \begin{bmatrix} \varepsilon(\boldsymbol{k}) + M(\boldsymbol{k}) & A_1 k_z & 0 & A_2 k_- \\ A_1 k_z & \varepsilon(\boldsymbol{k}) - M(\boldsymbol{k}) & A_2 k_- & 0 \\ 0 & A_2 k_+ & \varepsilon(\boldsymbol{k}) + M(\boldsymbol{k}) & -A_1 k_z \\ A_2 k_+ & 0 & -A_1 k_z & \varepsilon(\boldsymbol{k}) - M(\boldsymbol{k}) \end{bmatrix}, \tag{1.3}$$

where the reference energy is the middle of the bandgap, $\varepsilon(\boldsymbol{k}) = D_1 k_z^2 + D_2 k^2$, $M(\boldsymbol{k}) = -\Delta - B_1 k_z^2 - B_2 k^2$, $k_\pm = k_x \pm i k_y, k^2 = k_x^2 + k_y^2$. Parameters A, B, D can be determined from comparison to experimental results. Eigenvalues of the Hamiltonian (1.3) describe double spin degenerate conduction and valence bands:

$$E_{\mathrm{c,v}} = \varepsilon(\boldsymbol{k}) \pm \sqrt{M(\boldsymbol{k})^2 + A_1^2 k_z^2 + A_2^2 k^2}. \tag{1.4}$$

The symmetry of the Hamiltonian comprises the operations of the D_{3d} point group as well as the time reversal transformation. In the basis assumed in Hamiltonian (1.3), the representation of the time reversal (TR) and the spatial inversion (SI) operators are given by matrices

$$\mathrm{TR} = K \begin{pmatrix} 0 & 0 & 1 & 0 \\ 0 & 0 & 0 & 1 \\ -1 & 0 & 0 & 0 \\ 0 & -1 & 0 & 0 \end{pmatrix}, \quad \mathrm{SI} = \begin{pmatrix} 1 & 0 & 0 & 0 \\ 0 & -1 & 0 & 0 \\ 0 & 0 & 1 & 0 \\ 0 & 0 & 0 & -1 \end{pmatrix}, \tag{1.5}$$

where K stands for the complex conjugate. It is easy to check that the transformed Hamiltonians $(UHU^{+}, U = \mathrm{TR}, \mathrm{SI})$ have the same eigenvalues given in (1.4). Electron spectrum and band inversion are illustrated in Fig. 1.3.

The sign reversal of the gap parameter Δ corresponds to SOC-induced level crossing shown in Fig. 1.2. Band spectrum formation as shown in Fig. 1.3 relies on the nonzero momentum matrix element $A_2 \sim \langle P1_z^+ \uparrow |\nabla_{x,y}| P2_z^- \downarrow\rangle = \langle P1_z^+ \downarrow |\nabla_{x,y}| P2_z^- \uparrow\rangle$

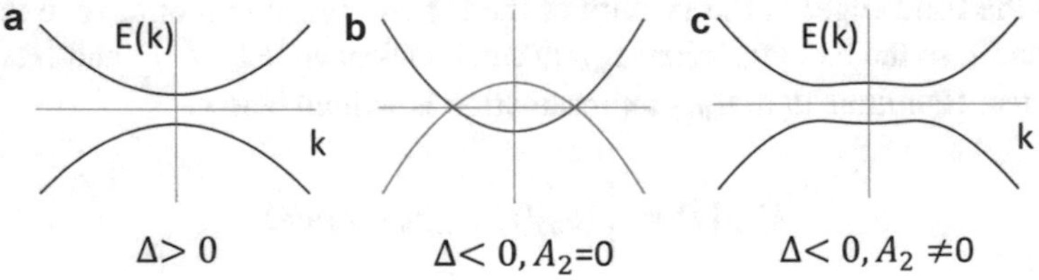

Fig. 1.3 Formation of the band inverted spectrum (1.4) at $k_z = 0$, $B_2 > 0$

which lifts degeneracy and results in the camelback shape of the inverted energy bands illustrated in Fig. 1.3c. This matrix element also stems from SOC and may exist only if wave functions have opposite parities. Inverted bands also take place in the z-direction if $\Delta B_1 < 0$. The band inversion creates the non-trivial topology of the wave functions that determines the main characteristics of TI solids. As discussed in the next section, the surface normal to the z-axis carries topologically non-trivial electron states if the k_z-bands are inverted.

1.3 Surface States

We start with the Hamiltonian (1.3) at $D_1 = D_2 = 0$. This simplification implies equal masses of electrons and holes that is not essential in the study of topological surface states. We account for the finite size of the sample in the z-direction, making the substitution, $k_z \to -i\nabla_z$. For the moment, let us restrict our consideration to the surface level at $k = 0$ and deal with the Hamiltonian

$$H_0 = \begin{pmatrix} -\Delta + B_1\frac{\partial^2}{\partial z^2} & -iA_1\frac{\partial}{\partial z} & 0 & 0 \\ -iA_1\frac{\partial}{\partial z} & \Delta - B_1\frac{\partial^2}{\partial z^2} & 0 & 0 \\ 0 & 0 & -\Delta + B_1\frac{\partial^2}{\partial z^2} & iA_1\frac{\partial}{\partial z} \\ 0 & 0 & iA_1\frac{\partial}{\partial z} & \Delta - B_1\frac{\partial^2}{\partial z^2} \end{pmatrix}. \tag{1.6}$$

Hamiltonian H_0 is a block-diagonal matrix with upper and lower blocks corresponding to spin-up and spin-down electrons, respectively. It is sufficient to consider the upper block only and then use the resulting wave functions substituting $A_1 \to -A_1$ to get the characteristics of the lower block. So, we will be dealing with a 2×2 Hamiltonian that describes spin-up electrons and holes:

$$H_0 = \begin{pmatrix} -\Delta + B_1\frac{\partial^2}{\partial z^2} & -iA_1\frac{\partial}{\partial z} \\ -iA_1\frac{\partial}{\partial z} & \Delta - B_1\frac{\partial^2}{\partial z^2} \end{pmatrix}. \tag{1.7}$$

Electronic states near the surface are in-plane Bloch waves while in the z-direction they can be described by a z-dependent spinor: $\Psi_\uparrow = A\begin{pmatrix} |1\rangle \\ |2\rangle \end{pmatrix}\exp(\lambda z)$. From the Schrödinger equation $H_0\Psi_\uparrow = E\Psi_\uparrow$, we get the characteristic equation for λ:

$$\mathrm{Det}\begin{pmatrix} -\Delta + B_1\lambda^2 - E & -iA_1\lambda \\ -iA_1\lambda & \Delta - B_1\lambda^2 - E \end{pmatrix} = 0. \tag{1.8}$$

The equation has four solutions: :$\pm\lambda_1, \pm\lambda_2$:

$$\lambda_{1,2} = \frac{1}{B_1\sqrt{2}}\left[F \mp \sqrt{F^2 + 4B_1^2(E^2 - \Delta^2)}\right]^{1/2},$$
$$F = A_1^2 + 2B_1\Delta. \tag{1.9}$$

Real $\lambda_{1,\,2}$ guarantee the exponential decay of the wave function away from the surface and exist only in the energy interval inside the energy gap $|E| < |\Delta|$. Outside the gap, the imaginery part of λ makes the wave function oscillatory and merges with the bulk states. For an electron energy that falls within the gap, additional constrains following from (1.9) ensure the existence of well-defined surface states:

$$B_1\Delta < 0; \quad A_1^2 > 4|B_1\Delta|. \tag{1.10}$$

So, surface states exist if the bulk energy spectrum in the k_z-direction is inverted. In order to find the energy spectrum of surface electrons, we present the wave function $\begin{pmatrix} |1\rangle \\ |2\rangle \end{pmatrix}$ as the superposition of two linearly independent eigenfunctions of the Hamiltonian (1.7) corresponding to real λ_1, λ_2. As we consider a semi-infinite sample with a single surface at $z = 0$, the wave function is expressed as:

$$\Psi_{\uparrow}(z) = C_1\begin{pmatrix} B_1\lambda_1^2 + E - \Delta \\ iA_1\lambda_1 \end{pmatrix}\exp(-\lambda_1 z)$$
$$+ C_2\begin{pmatrix} B_1\lambda_1^2 + E - \Delta \\ iA_1\lambda_2 \end{pmatrix}\exp(-\lambda_2 z). \tag{1.11}$$

The zero boundary condition $\Psi_{\uparrow}(z) = 0$ becomes a system of homogeneous equations for coefficients $C_{1,\,2}$:

$$X\begin{pmatrix} C_1 \\ C_2 \end{pmatrix} = 0,$$
$$X = \begin{pmatrix} B_1\lambda_1^2 + E - \Delta & B_1\lambda_2^2 + E - \Delta \\ iA_1\lambda_1 & iA_1\lambda_2 \end{pmatrix}. \tag{1.12}$$

The energy of surface state $E = 0$ compatible with real nonzero $\lambda_{1,\,2}$ follows as a solution to $\mathrm{Det}(X) = 0$:

$$(B_1\lambda_1^2 + E - \Delta)\lambda_2 - (B_1\lambda_2^2 + E - \Delta)\lambda_1 = 0. \tag{1.13}$$

One more solution $E = 0$ corresponds to $\Psi_{\downarrow}(z)$ and follows from the lower block of the Hamiltonian H_0. The exact $E = 0$ position of the surface level at $k = 0$ is the consequence of the electron-hole symmetry assumed in the Hamiltonian (1.3). If this condition were relaxed, the level would be shifted from the middle of the gap still located somewhere inside the gap of the bulk spectrum. It should be noted that the surface level pinned inside the bandgap discerns TI from conventional solids, where Tamm–Shockley surface states may appear due to the termination of the periodic potential at the surface. However, the existence and positions of Tamm–Shockley levels depend on the termination potential which may place them (if any) inside the conduction or the valence band. The level exists due to inverted bands, and its topological nature is manifested in the fact that the level is pinned to the bandgap of a bulk crystal irrespective of the surface potential and surface defects. The fact is related to the band topology and will be discussed in the next chapters.

Topological surface states are not unique to the TI solids specified above. The zero mode was first predicted in a heterostructure, where two IV–VI semiconductors with opposite signs of the gap parameter are brought into contact [3]. As will become clear later, the IV–VI single crystal also has surface states in the middle of the bulk gap energy region. This material class is called a crystalline topological insulator and includes the $Pb_{1-x}Sn_xTe(Se)$ system. The topological state is protected by mirror symmetry and occurs when Sn content exceeds the critical value at which conduction and valence bands invert their positions on an energy scale [4–7].

1.4 Thin Film

Below we consider the full TI model with two surfaces and look at surface modes at finite in-plane wave vector $k \neq 0$.

1.4.1 Wave Functions

We start with the Hamiltonian $H = H_0 + H_1$, where H_0 is given in (1.6) and

$$H_1 = [V_S(z) + V_{AS}(z)]I + \begin{pmatrix} -B_2k^2 & 0 & 0 & A_2k_- \\ 0 & B_2k^2 & A_2k_- & 0 \\ 0 & A_2k_+ & -B_2k^2 & 0 \\ A_2k_+ & 0 & 0 & B_2k^2 \end{pmatrix}, \tag{1.14}$$

where $I = \mathrm{diag}\,(1, 1, 1, 1)$ and the external field $V(z)$ may include band bending near surface and is written as the sum of $V_S(z) = [V(z) + V(-z)]/2$ and $V_{AS}(z) = [V(z) - V(-z)]/2$, even and odd parts, respectively. A nonzero $V_{AS}(z)$ implies spatial inversion asymmetry.

First, we consider the Hamiltonian (1.7) in a film with two surfaces at $z = \pm L/2$, so the wave function includes terms with $\pm\lambda_{1,\,2}$:

$$\Psi_\uparrow(z) = C_{11}\begin{pmatrix} B_1\lambda_1^2 + E - \Delta \\ -iA_1\lambda_1 \end{pmatrix}\exp(\lambda_1 z) + C_{12}\begin{pmatrix} B_1\lambda_1^2 + E - \Delta \\ iA_1\lambda_1 \end{pmatrix}\exp(-\lambda_1 z)$$
$$+ C_{21}\begin{pmatrix} B_1\lambda_2^2 + E - \Delta \\ -iA_1\lambda_2 \end{pmatrix}\exp(\lambda_2 z) + C_{22}\begin{pmatrix} B_1\lambda_2^2 + E - \Delta \\ iA_1\lambda_2 \end{pmatrix}\exp(-\lambda_2 z). \tag{1.15}$$

The problem has the spatial inversion symmetry: the surfaces at $z = \pm L/2$ are identical, so we can reduce the number of coefficients by assuming symmetric ($C_{11} = C_{12}, C_{21} = C_{22}$) and antisymmetric ($C_{11} = -C_{12}, C_{21} = -C_{22}$) superpositions in (1.15). Next we consider the two cases separately.

1. $C_{11} = C_{12} = A$, $C_{21} = C_{22} = B$. The wave function takes the form

$$\chi_\uparrow(z) = A\begin{pmatrix} (B_1\lambda_1^2 + E - \Delta)cosh[\lambda_1 z] \\ -iA_1\lambda_1 sinh[\lambda_1 z] \end{pmatrix} + B\begin{pmatrix} (B_1\lambda_2^2 + E - \Delta)cosh[\lambda_2 z] \\ iA_1\lambda_2 sinh[\lambda_2 z] \end{pmatrix}. \tag{1.16}$$

Zero boundary conditions $\chi_\uparrow(\pm L/2) = 0$ become a system of homogeneous equations for coefficients A, B:

$$M\begin{pmatrix} A \\ B \end{pmatrix} = 0,$$

$$M = A\begin{pmatrix} (B_1\lambda_1^2 + E - \Delta)\,cosh\,[\lambda_1 L/2] & (B_1\lambda_2^2 + E - \Delta)\,cosh\,[\lambda_2 L/2] \\ iA_1\lambda_1\,sinh\,[\lambda_1 L/2] & iA_1\lambda_2\,sinh\,[\lambda_2 L/2] \end{pmatrix}, \tag{1.17}$$

and the equation $\mathrm{Det}(M) = 0$ determines the surface energy level E_-:

$$\frac{(B_1\lambda_1^2 + E - \Delta)\lambda_2}{(B_1\lambda_2^2 + E - \Delta)\lambda_1} = \frac{tanh[\lambda_1 L/2]}{tanh[\lambda_2 L/2]}. \tag{1.18}$$

The wave function corresponding to this energy level can be found from (1.16) providing coefficients A, B are known. Using (1.17) and (1.18), one finds

$$B = -A\frac{B_1\lambda_1^2 + E - \Delta}{B_1\lambda_2^2 + E - \Delta}\cdot\frac{cosh[\lambda_1 L/2]}{cosh[\lambda_2 L/2]}. \tag{1.19}$$

Solving (1.18) with respect to $E - \Delta$, we get

$$E - \Delta + B_1\lambda_1^2 = \frac{B_1\lambda_1(\lambda_1^2 - \lambda_2^2)tanh[\lambda_1 L/2]}{\lambda_1 tanh[\lambda_1 L/2] - \lambda_2 tanh[\lambda_2 L/2]},$$

$$E - \Delta + B_1\lambda_2^2 = \frac{B_1\lambda_2(\lambda_2^2 - \lambda_1^2)tanh[\lambda_2 L/2]}{\lambda_2 tanh[\lambda_2 L/2] - \lambda_1 tanh[\lambda_1 L/2]}. \tag{1.20}$$

Now substituting the coefficient B from (1.19) into (1.16) and using relations (1.20), we come to an explicit form of the wave function corresponding to the energy level E_-:

$$\chi_\uparrow(z) = A\left[\begin{pmatrix} (B_1\lambda_1^2 + E - \Delta)\cosh[\lambda_1 z] \\ -iA_1\lambda_1 \sinh[\lambda_1 z] \end{pmatrix} - \frac{B_1\lambda_1^2 + E - \Delta}{B_1\lambda_2^2 + E - \Delta} \times \frac{\cosh[\lambda_1 L/2]}{\cosh[\lambda_2 L/2]} \times \begin{pmatrix} (B_1\lambda_2^2 + E - \Delta)\cosh[\lambda_2 z] \\ -iA_1\lambda_2 \sinh[\lambda_2 z] \end{pmatrix}\right] = N_\uparrow \begin{pmatrix} -B_1\eta_2(E_-)f_+(z, E_-) \\ iA_1 f_-(z, E_-) \end{pmatrix}, \tag{1.21}$$

where

$$\eta_2(E_-) = \frac{\lambda_1^2 - \lambda_2^2}{\lambda_1 tanh[\lambda_1 L/2] - \lambda_2 tanh[\lambda_2 L/2]},$$

$$f_+(z, E_-) = \frac{cosh[z\lambda_1]}{cosh[\lambda_1 L/2]} - \frac{cosh[z\lambda_2]}{cosh[\lambda_2 L/2]},$$

$$f_-(z, E_-) = \frac{sinh[z\lambda_1]}{sinh[\lambda_1 L/2]} - \frac{sinh[z\lambda_2]}{sinh[\lambda_2 L/2]}. \tag{1.22}$$

and $N_\uparrow$ is the normalization constant determined from the condition $\int_{-L/2}^{L/2} \chi_\uparrow^+ \chi_\uparrow dz = 1$.

2. $C_{11} = -C_{12} = A$, $C_{21} = -C_{22} = B$. The wave function that follows from (1.15) is expressed as

$$\varphi_\uparrow(z) = A\begin{pmatrix} (B_1\lambda_1^2 + E - \Delta)\sinh[\lambda_1 z] \\ -iA_1\lambda_1 \cosh[\lambda_1 z] \end{pmatrix} + B\begin{pmatrix} (B_1\lambda_2^2 + E - \Delta)\sinh[\lambda_2 z] \\ -iA_1\lambda_2 \cosh[\lambda_2 z] \end{pmatrix}. \tag{1.23}$$

Boundary conditions $\varphi_\uparrow(\pm L/2) = 0$ result in the equation for the surface state E_+:

$$\frac{(B_1\lambda_1^2 + E - \Delta)\lambda_2}{(B_1\lambda_2^2 + E - \Delta)\lambda_1} = \frac{tanh[\lambda_2 L/2]}{tanh[\lambda_1 L/2]}. \tag{1.24}$$

Following the procedure described after (1.18), we obtain an explicit form of the wave function corresponding to the energy level E_+:

$$\varphi_\uparrow(z) = C_\uparrow \begin{pmatrix} -B_1\eta_1(E_+)f_-(z,E_+) \\ iA_1 f_+(z,E_+) \end{pmatrix}, \tag{1.25}$$

where

$$\eta_1(E_+) = \frac{\lambda_1^2 - \lambda_2^2}{\lambda_1 coth[\lambda_1 L/2] - \lambda_2 coth[\lambda_2 L/2]},$$
$$f_+(z,E_+) = \frac{cosh[z\lambda_1]}{cosh[\lambda_1 L/2]} - \frac{cosh[z\lambda_2]}{cosh[\lambda_2 L/2]},$$
$$f_-(z,E_+) = \frac{sinh[z\lambda_1]}{sinh[\lambda_1 L/2]} - \frac{sinh[z\lambda_2]}{sinh[\lambda_2 L/2]}. \tag{1.26}$$

Wave functions $\varphi_\uparrow$ and $\chi_\uparrow$ correspond to energy levels E_+ and E_-, respectively, and present surface electron states at $k = 0$. The gap in the surface energy spectrum at Γ-point can be expressed analytically in the limit $\lambda_1 \gg \lambda_2$, $L\lambda_{1,\,2} \gg 1$:

$$\Delta_S = E_+ - E_- \approx 4\Delta \exp(-\lambda_2 L). \tag{1.27}$$

So, the zero mode $E = 0$ from (1.13) splits when the second surface is present and the spectrum becomes gapless again if the film thickness increases so that the overlap of wave functions localized near opposite surfaces tends to zero.

Two blocks in the block-diagonal Hamiltonian H_0 correspond to spin-up and spin-down electrons and they differ in the sign of the parameter A_1. So, wave functions $\varphi_\downarrow(z)$ and $\chi_\downarrow(z)$ can be obtained from spin-up functions $\varphi_\uparrow(z)$ and $\chi_\uparrow(z)$ by replacing $A_1 \rightarrow -A_1$. Energy levels $E_\pm$ do not change. This corresponds to the spin degeneracy of the spectrum. Four eigenstates of the Hamiltonian H_0 describe two surface levels with two spins and can be written as

$$\Phi_1(z) = \begin{pmatrix} \varphi_\uparrow(z) \\ 0 \end{pmatrix}, \Phi_2(z) = \begin{pmatrix} \chi_\uparrow(z) \\ 0 \end{pmatrix}, \Phi_3(z) = \begin{pmatrix} 0 \\ \varphi_\downarrow(z) \end{pmatrix}, \Phi_4(z) = \begin{pmatrix} 0 \\ \chi_\downarrow(z) \end{pmatrix}. \tag{1.28}$$

It is straightforward to check the orthogonality of the functions (1.28), so with the proper choice of normalization constants in (1.21), (1.25) we have

$$\int_{-L/2}^{L/2} \Phi_i^+(z)\Phi_j(z)dz = \delta_{ij}. \tag{1.29}$$

Orthogonality follows from the symmetry of the integrand. An example is shown below:

$$\Phi_1^+\Phi_2 = f_+(z,E_+)f_-(z,E_-)\left[A_1^2 + B_1^2\eta_1(E_+)\eta_2(E_-)\right]. \tag{1.30}$$

As it follows from (1.22) and (1.26), the product (1.30) is the odd function of z that gives zero after integration in the symmetric limits.

1.4.2 Effective Surface Hamiltonian

Using basis spinors (1.28), we compose the total wave function in a second quantization representation as follows:

$$\Psi(\boldsymbol{r}) = \frac{1}{\sqrt{V}}\sum_{\boldsymbol{k}}\sum_{i=1}^{4} a_{ik}\Phi_i(z)\exp(i\boldsymbol{kr}), \tag{1.31}$$

where V is the volume of the film, and $\boldsymbol{k}$ is the in-plane wave vector. Within the second quantization approach the wave function $\Psi(\boldsymbol{r})$ is considered to be an operator and coefficient $a_{ik}\left(a_{ik}^+\right)$ is the annihilation (creation) operator of the electrons in the basis state i with momentum $\boldsymbol{k}$. In this representation, the expectation value of energy presents the effective Hamiltonian acting in surface subspace [8]:

$$\tilde{H} = \int \Psi^+(\boldsymbol{r})H\Psi(\boldsymbol{r})d\boldsymbol{r}, \tag{1.32}$$

where H is given in (1.6) and (1.14):

$$\tilde{H} = \sum_{\boldsymbol{k}}\sum_{i,j=1}^{4}\tilde{H}_{ij}a_{ik}^+a_{ik}, \quad \tilde{H}_{ij} = \int_{-L/2}^{L/2}\Phi_i^+(z)H\Phi_j(z)dz. \tag{1.33}$$

The eigenvalues of $\tilde{H}_{ij}$ determine the surface energy spectrum in TI. As the basis functions were chosen as eigenfunctions of H_0, the diagonal matrix $\left(\tilde{H}_0\right)_{ij}$ comprises the surface energy levels on the main diagonal:

$$\left(\tilde{H}_0\right)_{ij} = \mathrm{diag}(E_+,E_-,E_+,E_-) = E_0 + \mathrm{diag}\left(\frac{\Delta_S}{2},-\frac{\Delta_S}{2},\frac{\Delta_S}{2},-\frac{\Delta_S}{2}\right),$$
$$E_0 = \frac{E_+ + E_-}{2}. \tag{1.34}$$

If the electron-hole symmetry holds, $E_0 = 0$. It is convenient to represent H_1 from (1.14) as 2×2 matrix:

$$H_1 = [V_S(z) + V_{AS}(z)]I + \begin{pmatrix} -B_2k^2\tau_z & A_2k_-\sigma_x \\ A_2k_+\sigma_x & -B_2k^2\tau_z \end{pmatrix}, \tag{1.35}$$

where τ_z and σ_x are the 2×2 Pauli matrices acting in two-band and two-spin space, respectively. The calculation of the matrix element $(\tilde{H}_1)_{14}$ is given below as an example:

$$\begin{aligned}(\tilde{H}_1)_{14} &= \langle \Phi_1|H_1|\Phi_4\rangle = \int_{-L/2}^{L/2} dz \left(\varphi_\uparrow^+(z), 0\right) \begin{pmatrix} -B_2k^2\tau_z & A_2k_-\sigma_x \\ A_2k_+\sigma_x & -B_2k^2\tau_z \end{pmatrix} \begin{pmatrix} 0 \\ \chi_\downarrow(z) \end{pmatrix} \\ &= A_2k_-\langle \varphi_\uparrow|\sigma_x|\chi_\downarrow\rangle. \end{aligned} \tag{1.36}$$

The matrix element $\langle \varphi_\uparrow|\sigma_x|\chi_\downarrow\rangle$ is imaginary

$$\langle \varphi_\uparrow|\sigma_x|\chi_\downarrow\rangle = iA_1 \int_{-L/2}^{L/2} \left[\eta_1(E_+)f_-(z,E_-)f_-(z,E_+) + \eta_2(E_-)f_+(z,E_-)f_+(z,E_+)\right] dz,$$

and (1.36) is finally expressed as

$$(\tilde{H}_1)_{14} = i\tilde{A}_2k_-, \quad \tilde{A}_2 \sim A_1A_2. \tag{1.37}$$

Direct calculation of all matrix elements gives an effective Hamiltonian in the form.

$$\begin{aligned}\tilde{H} &= \tilde{H}_0 + \tilde{H}_1 = (\tilde{V}_S + Dk^2)I \\ &+ \begin{pmatrix} \frac{\Delta_S}{2} - Bk^2 & \tilde{V}_{AS} & 0 & i\tilde{A}_2k_- \\ \tilde{V}_{AS} & -\frac{\Delta_S}{2} + Bk^2 & i\tilde{A}_2k_- & 0 \\ 0 & -i\tilde{A}_2k_+ & \frac{\Delta_S}{2} - Bk^2 & \tilde{V}_{AS} \\ -i\tilde{A}_2k_+ & 0 & \tilde{V}_{AS} & -\frac{\Delta_S}{2} + Bk^2 \end{pmatrix},\end{aligned} \tag{1.38}$$

where

$$D = \frac{\tilde{B}_1 + \tilde{B}_2}{2}, \quad B = \frac{\tilde{B}_1 - \tilde{B}_2}{2},$$

$$\tilde{B}_1 = B_2\langle\varphi_\uparrow|\tau_z|\varphi_\uparrow\rangle, \quad \tilde{B}_2 = B_2\langle\chi_\uparrow|\tau_z|\chi_\uparrow\rangle,$$
$$\tilde{V}_{AS} = \langle\varphi_\uparrow|V_{AS}|\chi_\uparrow\rangle = \langle\varphi_\downarrow|V_{AS}|\chi_\downarrow\rangle,$$
$$\tilde{V}_S = \langle\varphi_\uparrow|V_S|\varphi_\uparrow\rangle = \langle\chi_\uparrow|V_S|\chi_\uparrow\rangle. \tag{1.39}$$

Solutions to the equation $\mathrm{Det}\left(\tilde{H}_{ij} - E\right) = 0$ determine the four branches of the surface energy spectrum:

$$E_{\mathrm{c,v}\uparrow} = \tilde{V}_S + Dk^2 \pm R_\uparrow,$$
$$E_{\mathrm{c,v}\downarrow} = \tilde{V}_S + Dk^2 \pm R_\downarrow,$$
$$R_{\uparrow\downarrow} = \sqrt{\left(\frac{\Delta_S}{2} - Bk^2\right)^2 + \left(\tilde{A}_2 k \pm \tilde{V}_{AS}\right)^2}\,. \tag{1.40}$$

The signs $\pm$ in front of the square roots correspond to conduction and valence bands, and the signs under the roots stand for the spin variable, $\uparrow\downarrow$. Spin states $\uparrow\downarrow$ are the mixtures of pure spinors in the bulk. Spectrum (1.40) describes real surface electrons that can be experimentally probed with angle-resolved photoemission spectroscopy.

Surface bands (1.40) are spin-split if the spin–orbit interaction $\left(\tilde{A}_2 \neq 0\right)$ and the structural inversion asymmetry $\left(\tilde{V}_{AS} \neq 0\right)$ act simultaneously. This lifts the spin degeneracy so the bands remain Kramers degenerate if the time reversal symmetry holds, $E_\uparrow(\boldsymbol{k}) = E_\downarrow(-\boldsymbol{k})$.

The surface electrons behave like two-dimensional Rashba electron gas (for details on Rashba interaction in semiconductors see, for example, [9]). The spectrum is sensitive to the relative sign of the gap and the effective mass. If the surface spectrum has a direct gap ($B\Delta_S < 0$), the energy dispersion has the shape shown in Fig. 1.4.

If the surface bands are inverted ($B\Delta_S > 0$), the dispersion changes as illustrated in Fig. 1.5.

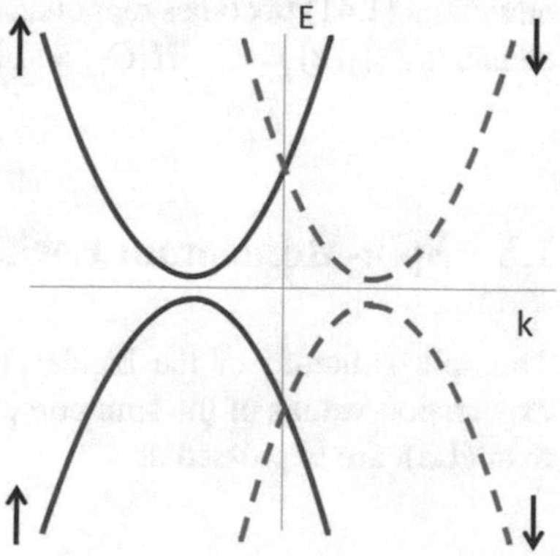

Fig. 1.4 Surface energy spectrum, direct tunneling gap, $B\Delta_S < 0$. Arrows indicate spin variables

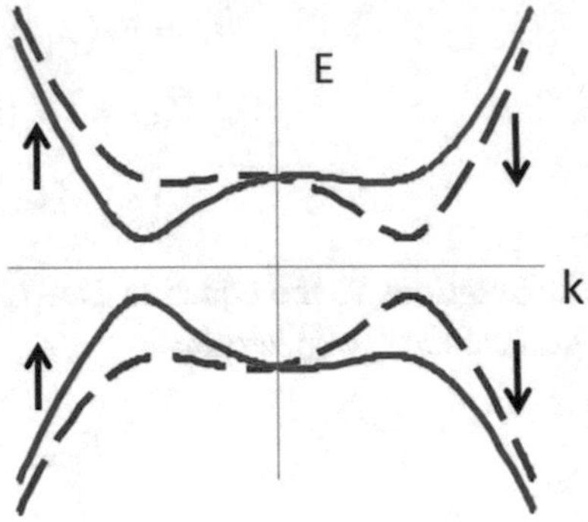

Fig. 1.5 Inverted band arrangement, $B\Delta_S > 0$. Dashed and solid lines illustrate branches with opposite spins

Eigenfunctions corresponding to energy bands (1.40) normalized to unity are given below:

$$C_{\uparrow\downarrow} = \sqrt{\frac{Bk^2 - \Delta_S/2 + R_{\uparrow\downarrow}}{4R_{\uparrow\downarrow}}} \begin{pmatrix} \dfrac{(\tilde{A}_2 k \mp \tilde{V}_{AS}) i \exp(-i\varphi)}{Bk^2 - \Delta_S/2 + R_{\uparrow\downarrow}} \\ \mp i \exp(-i\varphi) \\ \dfrac{\tilde{V}_{AS} \mp \tilde{A}_2 k}{Bk^2 - \Delta_S/2 + R_{\uparrow\downarrow}} \\ 1 \end{pmatrix},$$

$$V_{\uparrow\downarrow} = \sqrt{\frac{\Delta_S/2 - Bk^2 + R_{\uparrow\downarrow}}{4R_{\uparrow\downarrow}}} \begin{pmatrix} \dfrac{(\tilde{A}_2 k \mp \tilde{V}_{AS}) i \exp(-i\varphi)}{\Delta_S/2 - Bk^2 + R_{\uparrow\downarrow}} \\ \mp i \exp(-i\varphi) \\ \dfrac{\tilde{V}_{AS} \mp \tilde{A}_2 k}{\Delta_S/2 - Bk^2 + R_{\uparrow\downarrow}} \\ 1 \end{pmatrix}, \tag{1.41}$$

where $\tan \varphi = k_y/k_x$ and the upper (lower) sign corresponds to spin, $s = \uparrow, \downarrow$. It is straightforward to check that the spinors are orthonormal, $C_s^+ C_{s'} = V_s^+ V_{s'} = \delta_{ss'}; C_s^+ V_{s'} = 0$, and the matrix Hamiltonian calculated with the new basis (1.41) becomes diagonal. Energy bands (1.40) can be found as the matrix elements $E_{c\uparrow\downarrow}(k) = C_{\uparrow\downarrow}|\tilde{H}|C_{\uparrow\downarrow}$ and $E_{v,\uparrow\downarrow}(k) = V_{\uparrow\downarrow}|\tilde{H}|V_{\uparrow\downarrow}$.

1.5 Spin-Momentum Locked Fermions

The spin structure of the bands (1.40) can be found directly by calculating the expectation values of the spin components, which in the basis assumed in Hamiltonian (1.3), are expressed as

$$S_x = \begin{pmatrix} 0 & 0 & 1 & 0 \\ 0 & 0 & 0 & 1 \\ 1 & 0 & 0 & 0 \\ 0 & 1 & 0 & 0 \end{pmatrix}, \quad S_y = \begin{pmatrix} 0 & 0 & -i & 0 \\ 0 & 0 & 0 & -i \\ i & 0 & 0 & 0 \\ 0 & i & 0 & 0 \end{pmatrix}, \quad S_z = \begin{pmatrix} 1 & 0 & 0 & 0 \\ 0 & 1 & 0 & 0 \\ 0 & 0 & -1 & 0 \\ 0 & 0 & 0 & -1 \end{pmatrix}. \tag{1.42}$$

Spin expectation values calculated with spinors (1.41) are given as

$$\begin{aligned} \langle C_\uparrow | S_x | C_\uparrow \rangle &= -\sin\varphi, \quad \langle C_\uparrow | S_y | C_\uparrow \rangle = \cos\varphi, \quad \langle C_\uparrow | S_z | C_\uparrow \rangle = 0, \\ \langle C_\downarrow | S_x | C_\downarrow \rangle &= \sin\varphi, \quad \langle C_\downarrow | S_y | C_\downarrow \rangle = -\cos\varphi, \quad \langle C_\downarrow | S_z | C_\downarrow \rangle = 0. \end{aligned} \tag{1.43}$$

The result (1.43) is represented in Fig. 1.6.

Surface states with spin polarization spinning around the Dirac point as shown in Fig. 1.6 are the signature of a topological insulator. Their existence has been confirmed experimentally by angle-resolved photoemission and scanning tunneling microscopy techniques (for a review, see [10]). Backscattering is forbidden as it implies transition between spin-up and spin-down states that cannot occur unless a localized magnetic moment takes part in the process.

The energy bands (1.40) illustrated in Fig. 1.5 are shown with no indication of the surface they belong to. Electron localization near the top and bottom surfaces will be discussed below.

Let us consider the spatial dependence of the surface states when $k = 0$, $\tilde{V}_{AS} = 0$. The Hamiltonian (1.38) is diagonal and (1.28) represents its eigenfunctions. The electron probability density function $\Phi_1^+(z)\Phi_1(z)$ normalized to unity is shown in Fig. 1.7. The calculation uses the set of parameters given in Ref. [1]: $A_1 = 3.3$ eVÅ, $\Delta = 0.28$ eV, $B_1 = 1.5$ eVÅ^2.

Using the eigenstates (1.41), one can compose the z-dependent linear combination of basis wave functions $\Phi_{1-4}(z)$ which makes the Hamiltonian (1.38) diagonal with eigenvalues (1.40) on its main diagonal:

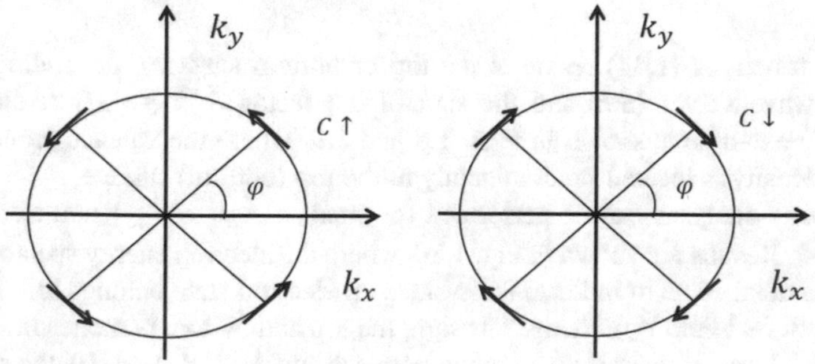

Fig. 1.6 Spin-momentum locked Dirac electrons

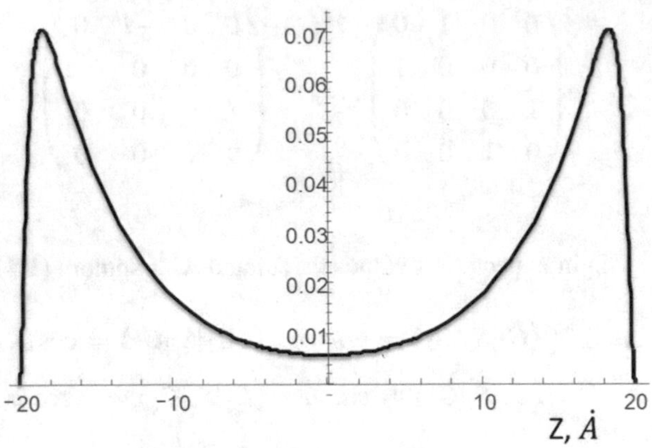

Fig. 1.7 Electron probability density $\Phi_1^+(z)\Phi_1(z)$ across a 40 Å-thick TI film

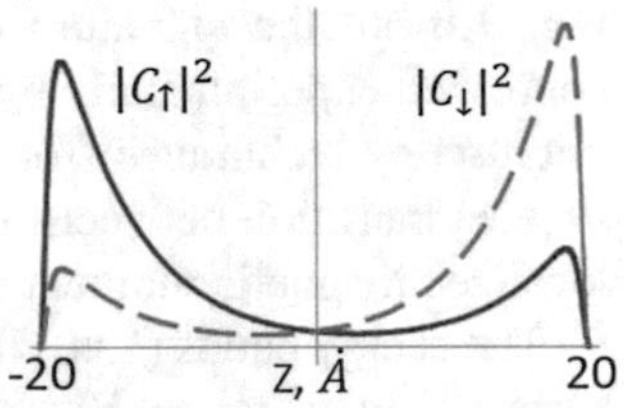

Fig. 1.8 Conduction band probability density, $(\tilde{A}_2k > \tilde{V}_{AS})$. Spin-up solid line, spin-down dashed line

$$\begin{aligned}
V_\uparrow(z) &= \frac{(\tilde{A}_2k - \tilde{V}_{AS})i\exp(-i\varphi)}{\Delta_S/2 - Bk^2 + R_\uparrow}\Phi_1(z) - i\exp(-i\varphi)\Phi_2(z) + \frac{\tilde{A}_2k - \tilde{V}_{AS}}{\Delta_S/2 - Bk^2 + R_\uparrow}\Phi_3(z) + \Phi_4(z),\\
C_\uparrow(z) &= \frac{(\tilde{A}_2k - \tilde{V}_{AS})i\exp(-i\varphi)}{Bk^2 - \Delta_S/2 + R_\uparrow}\Phi_1(z) - i\exp(-i\varphi)\Phi_2(z) + \frac{\tilde{V}_{AS} - \tilde{A}_2k}{Bk^2 - \Delta_S/2 + R_\uparrow}\Phi_3(z) + \Phi_4(z),\\
V_\downarrow(z) &= -\frac{(\tilde{A}_2k + \tilde{V}_{AS})i\exp(-i\varphi)}{\Delta_S/2 - Bk^2 + R_\downarrow}\Phi_1(z) + i\exp(-i\varphi)\Phi_2(z) - \frac{\tilde{A}_2k + \tilde{V}_{AS}}{\Delta_S/2 - Bk^2 + R_\downarrow}\Phi_3(z) + \Phi_4(z),\\
C_\downarrow(z) &= \frac{(\tilde{A}_2k + \tilde{V}_{AS})i\exp(-i\varphi)}{Bk^2 - \Delta_S/2 + R_\downarrow}\Phi_1(z) + i\exp(-i\varphi)\Phi_2(z) + \frac{\tilde{V}_{AS} + \tilde{A}_2k}{Bk^2 - \Delta_S/2 + R_\downarrow}\Phi_3(z) + \Phi_4(z).
\end{aligned} \tag{1.44}$$

Wave functions (1.44) reside at the top or bottom surfaces, depending on the in-plane wave vector ($\pm k$) and the sign of the factor $\tilde{A}_2k + \tilde{V}_{AS}$. A direct gap ($B\Delta_S < 0$) example is shown in Figs. 1.8 and 1.9, where the valence (conduction) electron density is located predominantly at the top (bottom) surface.

The same analyses can be performed for small $k\left(\tilde{A}_2k < \tilde{V}_{AS}\right)$ and also for the states at $-k$. Results are shown in Fig. 1.10, where the electron energy spectrum from Fig. 1.4 is detailed as to indicate the surface an electron state belongs to.

If an external field $\tilde{V}_{AS}$ changes its sign, the top and bottom branches in Fig. 1.10 swap their places. If there is no coupling between surfaces ($B, \Delta_S \to 0$), the spectrum

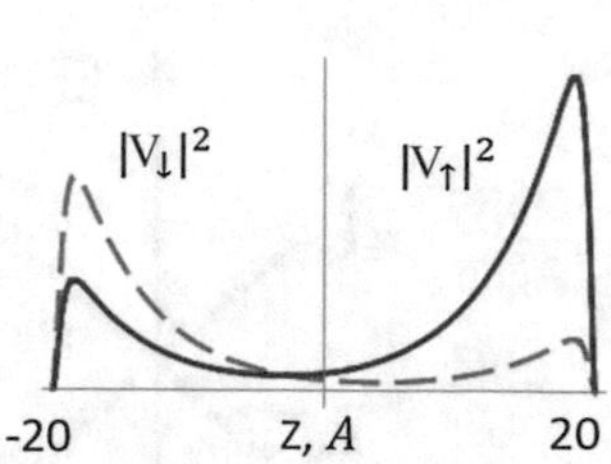

Fig. 1.9 Valence band probability density, $(\tilde{A}_2 k > \tilde{V}_{AS})$. Spin-up solid line, spin-down dashed line

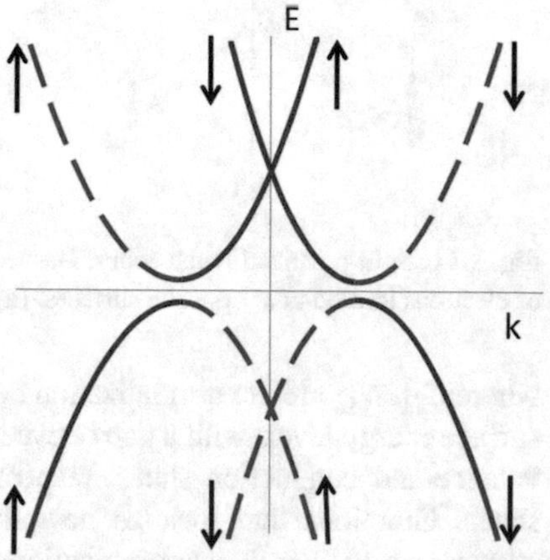

Fig. 1.10 Electron spectrum (1.40), $B\Delta_S < 0$. Solid (bottom) and dashed (top) lines indicate electron localization near opposite surfaces

(1.40) describes the pair of massless Dirac cones residing on opposite surfaces and separated in the energy scale by $2\tilde{V}_{AS}$:

$$E_{c,v,\uparrow,\downarrow} = V_S \pm \left(\tilde{A}_2 k \pm \tilde{V}_{AS}\right). \tag{1.45}$$

The spectrum (1.45) is shown in Fig. 1.11.

If the vertical bias is absent, the surface Dirac cones are degenerate as shown in Fig. 1.11b. The top-bottom tunneling opens the gap at $E(k) = 0$ transforming the spectrum in Fig. 1.11a into that shown in Fig. 1.10.

1.6 Top-Bottom Representation

Surface Hamiltonian (1.38) is given in the representation of the bases functions (1.28) which are written below as 4-spinors:

$$\begin{aligned}
\Phi_1^T &= C_\uparrow\left[-B_1\eta_1(E_+)f_-(z,E_+), iA_1f_+(z,E_+), 0, 0\right],\\
\Phi_2^T &= N_\uparrow\left[-B_1\eta_2(E_-)f_+(z,E_-), iA_1f_-(z,E_-), 0, 0\right],\\
\Phi_3^T &= C_\downarrow\left[0, 0, -B_1\eta_1(E_+)f_-(z,E_+), -iA_1f_+(z,E_+)\right],\\
\Phi_4^T &= N_\downarrow\left[0, 0, -\mathrm{B}_1\eta_2(E_-)f_+(z,E_-), -iA_1f_-(z,E_-)\right],
\end{aligned} \tag{1.46}$$

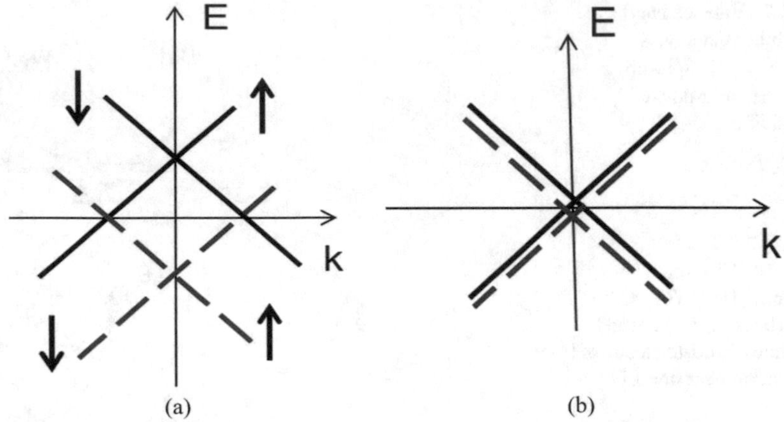

Fig. 1.11 Spin-polarized Dirac cones. Dashed and solid lines differentiate the energy dispersions of electrons located near opposite surfaces. (**a**) $\tilde{V}_{AS} \neq 0$. (b) $\tilde{V}_{AS} \to 0$

where $C_{\uparrow\downarrow}$, $N_{\uparrow\downarrow}$ are the normalization constants. Since wave functions correspond to surface energy levels with a gap between them, we refer to Φ_1, Φ_2 as spin-up surface valence and conduction states, respectively. Similarly, Φ_3, Φ_4 denote spin-down states. Functions that localize near the top (t) and bottom (b) surfaces can be represented as their linear combinations. Transition to the top-bottom representation is carried out by a unitary transformation

$$\begin{pmatrix} \mathrm{b}_\uparrow \\ \mathrm{b}_\downarrow \\ \mathrm{t}_\uparrow \\ \mathrm{t}_\downarrow \end{pmatrix} = U \begin{pmatrix} \Phi_1 \\ \Phi_2 \\ \Phi_3 \\ \Phi_4 \end{pmatrix},$$

$$U = \frac{1}{\sqrt{2}} \begin{pmatrix} 1 & 1 & 0 & 0 \\ 0 & 0 & 1 & 1 \\ -1 & 1 & 0 & 0 \\ 0 & 0 & -1 & 1 \end{pmatrix}, \quad |\mathrm{Det}[U]| = 1. \tag{1.47}$$

Hamiltonian (1.38) in the top-bottom representation is given as

$$H_{\mathrm{tb}} = U\tilde{H}U^{-1} = \begin{pmatrix} \tilde{V}_{AS} & i\tilde{A}_2 k_- & -\frac{\Delta_S}{2} + Bk^2 & 0 \\ -i\tilde{A}_2 k_+ & \tilde{V}_{AS} & 0 & -\frac{\Delta_S}{2} + Bk^2 \\ -\frac{\Delta_S}{2} + Bk^2 & 0 & -\tilde{V}_{AS} & -i\tilde{A}_2 k_- \\ 0 & -\frac{\Delta_S}{2} + Bk^2 & i\tilde{A}_2 k_+ & -\tilde{V}_{AS} \end{pmatrix}. \tag{1.48}$$

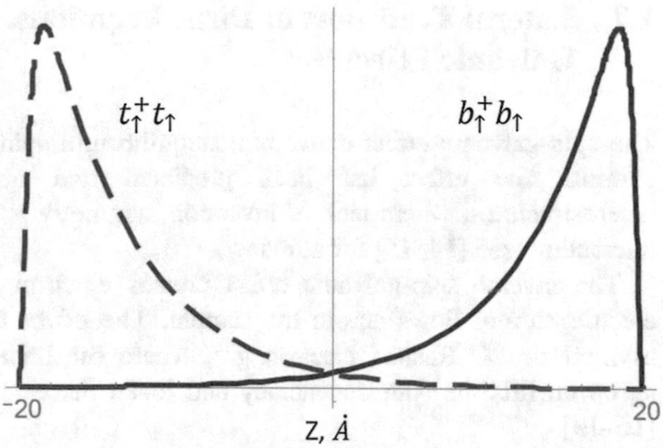

Fig. 1.12 Top and bottom electrons in a 40 Å-thick film. The positive z-direction points down to the bottom surface

At $\tilde{V}_{AS} = 0$, the matrix (1.48) rearranged in an appropriate sequence of bases functions, coincides with the Hamiltonian used in [11]. Probability densities are shown in Fig. 1.12.

In a thick film where top-bottom tunneling is negligible ($\Delta_S, B \to 0$), the Hamiltonian (1.48) is a block-diagonal where the upper and lower 2×2 blocks represent electron energy on the bottom and top surfaces, respectively. The Hamiltonian (1.48) can be expressed using Pauli matrices in the spin and top-bottom spaces, $\boldsymbol{\sigma}$ and $\boldsymbol{\tau}$, respectively:

$$H_{\text{tb}} = m_z \tau_0 \otimes \sigma_z + \tilde{V}_{AS} \tau_z \otimes \sigma_0 + \left(Bk^2 - \frac{\Delta_S}{2} \right) \tau_x \otimes \sigma_0 + \tilde{A}_2 \tau_z \otimes [\boldsymbol{\sigma} \times \boldsymbol{k}]_z, \quad (1.49)$$

where $\otimes$ means the Kronecker product and τ_0 and σ_0 are 2×2 unit matrices. The top-bottom model (1.49) in a simplified form has been used to study the Hall conductivity in QAHE [11], the QHE in the perpendicular magnetic field [12], and the quantum phase transition driven by the in-plane magnetic field [13]. In Chap. 3, we discuss a thin TI slab in a magnetic field using the top-bottom representation that makes calculations more transparent and easier to understand.

Using a particular semiconductor as an example, we have described a property common to the whole class of TI materials: the existence of edge zero modes. Spin–orbit coupling makes these modes helical in nature, that is, counterpropagating massless fermions carry opposite spins. The fact that spin is locked to momentum precludes elastic electron backscattering because this would require a change of the spin state during the scattering of an electron on a non-magnetic impurity.

This makes surface excitations robust against scattering on non-magnetic defects.

1.7 Lateral Transport of Dirac Fermions. Inverse Spin-Galvanic Effect

The spin-galvanic effect converts nonequilibrium spin polarization into electrical current. The effect has been predicted and observed in semiconductor heterostructures, where lack of inversion symmetry is accompanied by spin–orbit interaction (see [14, 15] for a review).

The inverse spin-galvanic effect creates electron spin polarization when an electric current flows across the sample. The effect takes place in an inversion asymmetric 2D Rashba electron gas, where the linear in $\boldsymbol{k}$ term of the energy spectrum lifts the spin degeneracy and locks the electron momentum to its spin [16–19].

A single helical surface of a TI solid presents another object, where the inverse spin-galvanic effect takes place: while in thermodynamic equilibrium the spin polarization of surface electrons is absent, the lateral ohmic electric current flowing on a gapless (Dirac-type) surface induces nonequilibrium spin polarization. Lateral current on a single surface presents a real experimental setting in a 3D TI solid which can be described by one of diagonal blocks in (1.48) under the assumption that the top-bottom mixing terms and vertical bias are negligible, $\Delta_S, B, \tilde{V}_{AS} \to 0$:

$$H = \begin{pmatrix} 0 & i\tilde{A}_2 k_- \\ -i\tilde{A}_2 k_+ & 0 \end{pmatrix}. \tag{1.50}$$

Branches of linear dispersion $\pm E_{\boldsymbol{k}}, E_{\boldsymbol{k}} = \tilde{A}\,_2 k$ describe helical modes of a single Dirac cone.

Magnetic moment per unit square (magnetization) of surface electrons can be calculated from the Hamiltonian (1.50) as follows:

$$\boldsymbol{m} = \frac{\mu_B T}{S} \sum_{\boldsymbol{k}} \sum_{i\omega_n} Tr[\boldsymbol{\sigma} G(i\omega_n)],$$

$$G(i\omega_n) = (i\omega_n + \mu - H)^{-1} = \frac{1}{(i\omega_n + \mu)^2 - E_k^2} \begin{pmatrix} i\omega_n + \mu & i\tilde{A}_2 k_- \\ -i\tilde{A}_2 k_+ & i\omega_n + \mu \end{pmatrix}, \tag{1.51}$$

where μ_B is the Bohr magneton, S is the surface area, $G(i\omega_n)$ is the Matsubara Green function [20, 21], and μ is the chemical potential. Trace calculation gives

$$m_x \sim \frac{2\tilde{A}_2 k_y}{(i\omega_n)^2 - E_k^2}; \quad m_y \sim \frac{-2\tilde{A}_2 k_x}{(i\omega_n)^2 - E_k^2}; \quad m_z = 0. \tag{1.52}$$

The Matsubara frequency sum in (1.51) is calculated using the relation

$$T\sum\nolimits_{\omega_n} g(i\omega_n) = -\frac{1}{2}\sum\nolimits_j \mathrm{Res}\left\{g(z_j)\, tanh\left[\frac{z_j}{2T}\right]\right\}, \tag{1.53}$$

where the residues are taken in z_j, the poles of $g(z)$ on the complex plane $z = i\omega_n$.

Calculations in (1.51) give

$$\frac{m_y}{\mu_B} = \frac{1}{(2\pi)^2}\int_0^\infty kdk \int_0^{2\pi} [f(E_{\boldsymbol{k}} - \mu) - f(-E_{\boldsymbol{k}} - \mu)]\ \cos\varphi\, d\varphi, \tag{1.54}$$

where $f(\varepsilon)$ is the electron distribution function. In the m_x component, $cos\varphi$ in (1.54) should be replaced with $sin\varphi$. In thermodynamic equilibrium, the distribution function $f(\varepsilon)$ should be replaced with the Fermi function f_0 in (1.54) that turns the magnetization to zero, so in equilibrium, the Kramers degenerate electron spectrum precludes magnetization from appearing.

In an electric field E_x, the distribution function is the solution of the Boltzmann equation, which in the relaxation time approximation is given as $f(\varepsilon) = f_0 - eE_x v_x \tau\ \partial f_0/\partial\varepsilon$, where $v_x = \tilde{A}_2 cos\varphi/\hbar$. Assuming the Fermi level lies above the Dirac point and $f'(\varepsilon) \to -\delta(\varepsilon)$, $T \to 0$, we arrive at

$$\frac{m_y}{\mu_B} = -\frac{eE_x\,\tau\,\tilde{A}_2}{(2\pi)^2\hbar}\int_0^\infty kdk \int_0^{2\pi} f'(E_{\boldsymbol{k}} - \mu)\ \cos^2\varphi \mathrm{d}\varphi = \frac{eE_x\tau k_F}{4\pi\hbar}. \tag{1.55}$$

The electric field E_x drives electric current density

$$\begin{aligned} J_x &= \frac{e}{S}\sum_{k_x,k_y} v_x f(\boldsymbol{k}) = \frac{e\tilde{A}_2}{(2\pi)^2\,\hbar}\int_0^\infty kdk \int_0^{2\pi} f(E_{\boldsymbol{k}} - \mu)\cos\varphi\, d\varphi \\ &= -\frac{e^2 E_x \tilde{A}_2^2 \tau}{(2\pi)^2\,\hbar^2}\int_0^\infty kdk \int_0^{2\pi} f'(E_{\boldsymbol{k}} - \mu)\cos^2\varphi\, d\varphi = \frac{e^2\tau\, E_x\tilde{A}_2 k_F}{4\pi\hbar^2}. \end{aligned} \tag{1.56}$$

Comparing (1.55) and (1.56), we arrive at current-induced magnetization as

$$m_y = \frac{\mu_B\hbar}{e\tilde{A}_2}J_x. \tag{1.57}$$

Expression (1.57) is obtained for a single Dirac cone of massless electrons residing on a surface. In a film, the situation is different as one has to account for a second cone residing on the opposite surface. The spectrum in the vertically unbiased film, shown in Fig. 1.11b, is spin degenerate turning the magnetic moment to zero at each $\boldsymbol{k}$-point, so, no matter whether current flows or not, the magnetization is zero if contributions from both surfaces are taken into account. These qualitative considerations can be confirmed by microscopic calculations of electron magnetic moment following from the block-diagonal 4 × 4 Hamiltonian (1.48) in the limit of independent surfaces, Δ_S, $B \to 0$:

$$H_f = \begin{pmatrix} \tilde{V}_{AS} & i\tilde{A}_2 k_- & 0 & 0 \\ -i\tilde{A}_2 k_+ & \tilde{V}_{AS} & 0 & 0 \\ 0 & 0 & -\tilde{V}_{AS} & -i\tilde{A}_2 k_- \\ 0 & 0 & i\tilde{A}_2 k_+ & -\tilde{V}_{AS} \end{pmatrix}. \tag{1.58}$$

Trace operation in (1.51) takes the form

$$\boldsymbol{m} \sim Tr\left[\boldsymbol{\Sigma} G_f(i\omega_n)\right], \quad \boldsymbol{\Sigma} = \begin{pmatrix} \boldsymbol{\sigma} & 0 \\ 0 & \boldsymbol{\sigma} \end{pmatrix}, \quad G_f(i\omega_n) = \left(i\omega_n + \mu - H_f\right)^{-1}.$$

$$\boldsymbol{m}_{x,y} \sim \frac{2\tilde{A}_2 k_{y,x}\left[\left(\tilde{V}_{AS} + i\omega_n\right)^2 - \left(\tilde{V}_{AS} - i\omega_n\right)^2\right]}{\left[\tilde{A}^2 k^2 - \left(\tilde{V}_{AS} + i\omega_n\right)^2\right]\left[\tilde{A}^2 k^2 - \left(\tilde{V}_{AS} - i\omega_n\right)^2\right]}. \tag{1.59}$$

In an unbiased film ($\tilde{V}_{AS} = 0$), Dirac cones are spin degenerate due to both TR and SI symmetries, and the magnetization (1.59) is identically zero. Vertical bias breaks spatial inversion symmetry; the electron spectrum of the film becomes Kramers degenerate (see Fig. 1.11a) that result in voltage-controlled current-induced magnetization. In a real experimental setting, built-in vertical bias always exists in a film deposited onto a substrate. Then, a lateral current induces electron magnetization signaling the inverse spin-galvanic effect. As for the direct spin-galvanic effect, that is the galvanic current induced by a spin flow [22], measurements imply the spin injection across the ferromagnetic material (FM)/sample interface and the current signal can be masked by the Hall signal originating from FM stray fields. Demonstration of the direct effect has been reported making use the special technique which separates the FM source from the TI sample by placing the *Si*-spacer between them and injecting spin current through the Si/Bi_2Se_3 Schottky barrier [23].

Potentiometric detection of electron spin polarization on a single Bi_2Se_3 surface, proposed in [24–26], has been performed with the experimental setup that includes ferromagnetic probe contact as shown in Fig. 1.13.

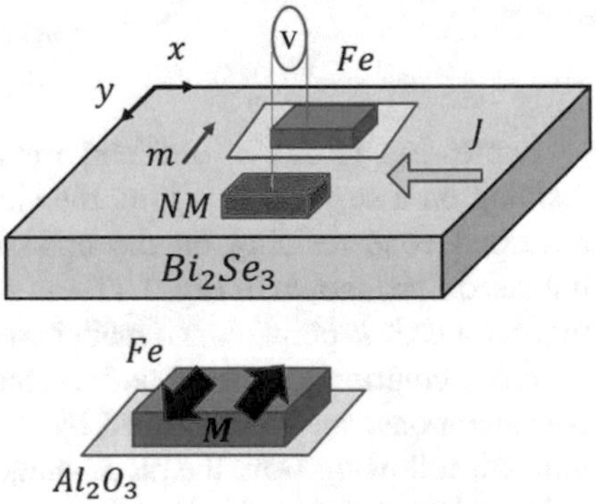

Fig. 1.13 Experimental setup for demonstration of the inverse spin-galvanic effect from surface Dirac fermions [26, 27]. Change in magnetization orientation in *Fe* detector contact under applied magnetic field generates the signal $V(M) - V(-M)$, proportional to the current-induced electron magnetic moment *m*

Spin to charge current conversion in $Bi_{2-x}Sb_xTe_{3-y}Se_y$ has been reported in [28]. Device-oriented spintronic applications imply an operation of the TI-based device as a room temperature spin injector when interfaced with other semiconductors. The room temperature inverse spin-galvanic effect has been measured in Bi_2Se_3 [29] and $\alpha - Sn$ [30].

The microscopic approach used in this chapter deals with the energy spectrum and reveals some features of TI solids. However, up to this point, we have not discussed the topological response to external perturbations despite the fact that the topological phase can be identified by that response. In this regard, a variety of Hall effects will be briefly discussed in the next chapters.

References

1. Zhang, H., Liu, C.-X., Qi, X.-L., Dai, X., Fang, Z., Zhang, S.-C.: Nat. Phys. **5**, 438–442 (2009)
2. Liu, C.-X., Qi, X.-L., Zhang, H., Dai, X., Fang, Z., Zhang, S.-C.: Phys. Rev. B. **82**, 045122 (2010)
3. Volkov, B.A., Pankratov, O.A.: JETP Lett. **42**, 178–181 (1985)
4. Fu, L.: Phys. Rev. Lett. **106**, 106802 (2011)
5. Tanaka, Y., Ren, Z., Sato, T., Nakayama, K., Souma, S., Takahashi, T., Segawa, K., Ando, Y.: Nat. Phys. **8**, 800–803 (2012)
6. Hsieh, T.H., Lin, H., Liu, J., Duan, W., Bansil, A., Fu, L.: Nat. Commun. **3**, 982 (2012)
7. Dziawa, P., Kowalski, B.J., Dybko, K., Buczko, R., Szczerbakow, A., Szot, M., Łusakowska, E., Balasubramanian, T., Wojek, B.M., Berntsen, M.H., Tjernberg, O., Story, T.: Nat. Mater. **11**, 1023–1027 (2012)
8. Shan, W.Y., Lu, H.Z., Shen, S.Q.: New J. Phys. **12**, 043048 (2010)
9. Litvinov, V.: Wideband Semiconductor Spintronics. Pan Stanford, Singapore (2016)
10. Hasan, M.Z., Kane, C.L.: Rev. Mod. Phys. **82**, 3045–3067 (2010)
11. Yu, R., Zhang, W., Zhang, H.J., Zhang, S.C., Dai, X., Fang, Z.: Science. **329**, 61 (2010)
12. Zyuzin, A.A., Burkov, A.A.: Phys. Rev. **B83**, 195413 (2011)
13. Zyuzin, A.A., Hook, M.D., Burkov, A.A.: Phys. Rev. **B83**, 245428 (2011)
14. Ivchenko, E., Ganichev, S.D.: arXiv preprint, arXiv:1710.09223 (2017)
15. Ganichev, S.D., Trushin, M., Schliemann, J.: Handbook of Spin Transport and Magnetism. Chapman and Hall, London (2016)
16. Edelstein, V.M.: Solid State Commun. **73**, 233 (1990)
17. Manchon, A., Zhang, S.: Phys. Rev. **B78**, 212405 (2008)
18. Miron, I.M., Gaudin, G., Auffret, S., Rodmacq, B., Schuhl, A., Pizzini, S., Vogel, J., Gambardella, P.: Nat. Mater. **9**, 230 (2010)
19. Dyrdał, J.B.'s., Dugaev, V.K.: Phys. Rev. **B95**, 245302 (2017)
20. Abrikosov, A., Gor'kov, L., Dzyaloshinskii, I.: Methods of Quantum Field Theory in Statistical Physics. Dover Publ, New York (1975)
21. Mahan, G.D.: Many-Particle Physics, 3rd edn. Kulwer Academic/Plenum, New York (2000)
22. Ganichev, S.D., Ivchenko, E.L., Bel'kov, V.V., Tarasenko, S.A., Sollinger, M., Weiss, D., Wegscheider, W., Prettl, W.: Nature. **417**, 153 (2002)
23. Ojeda-Aristizabal, C., Fuhrer, M.S., Butch, N.P., Paglione, J., Appelbaum, I.: Appl. Phys. Lett. **101**, 023102 (2012)
24. Silsbee, R.H.: J. Phys. Condens. Matter. **16**, R179 (2004)
25. Hong, S., Diep, V., Datta, S., Chen, K.P.: Phys. Rev. **B86**, 085131 (2012)
26. Li, C.H., Erve, O.M.J.v.'t., Rajput, S., Li, L., Jonker, B.T.: Nat. Commun. **7**, 13518 (2016)

27. Tang, J., Chang, L.-T., Kou, X., Murata, K., Choi, E.S., Lang, M., Fan, Y., Jiang, Y., Montazeri, M., Jiang, W., Wang, Y., He, L., Wang, K.L.: Nano Lett. **14**, 5423 (2014)
28. Shiomi, Y., Nomura, K., Kajiwara, Y., Eto, K., Novak, M., Segawa, K., Ando, Y., Saitoh, E.: Phys. Rev. Lett. **113**, 196601 (2014)
29. Dankert, A., Geurs, J., Kamalakar, M.V., Charpentier, S., Dash, S.P.: Nano Lett. **15**, 7976 (2015)
30. Rojas-Sánches, J.-C., Oyarzún, S., Fu, Y., Marty, A., Vergnaud, C., Gambarelli, S., Vila, L., Jamet, M., Ohtsubo, Y., Taleb-Ibrahimi, A., Le Fèvre, P., Bertran, F., Reyren, N., George, J.-M., Fert, A.: Phys. Rev. Lett. **116**, 096602 (2016)

Chapter 2
Hall Effects and Berry Phase

The classification of topological phases requires parameters that differentiate TI from conventional dielectrics. Solid state phases that differ in crystalline symmetry can be characterized by the order parameter. That's what the Landau theory of phase transitions deals with. The onset of the nonzero order parameter signals the breaking of symmetry and the transition to a low-symmetry phase when temperature or other system parameters change. TI cannot be described this way as no symmetry breaking occurs at the topological phase transition and thus no order parameter exists. The topologically non-trivial phase can be distinguished by topological invariants, the numbers that characterize the topology of the Hilbert space, in other words, certain properties of electron wave functions as they flow during the movement of electron momentum across the first Brillouin zone (BZ). The state with a nonzero topological invariant reveals unusual properties of a crystal boundary where trivial and non-trivial regions touch each other: backscattering is forbidden for electrons pinned to the interface (edge, surface). This protects topological interfaces against electron scattering on crystal imperfections. An example topological interface is the edge of a two-dimensional Quantum Hall Effect (QHE) system where the time reversal symmetry is broken by a perpendicular magnetic field and the Chern number plays the role of a topological invariant that protects the chiral electron edge states. Another example is the TI surface, also called Quantum Spin Hall Effect (QSHE) system, where time reversal symmetry holds and spin-resolved counterpropagating surface states (see Chap. 1) are protected by the Z_2 invariant [1]. Below we discuss various Hall effects and their common features related to the topological properties of the energy bands.

V. Litvinov, *Magnetism in Topological Insulators*,
https://doi.org/10.1007/978-3-030-12053-5_2

2.1 Hall Effects

The brief discussion of Hall effects in this section is preliminary in nature and concerns phenomenological relations. More details will be given throughout the text. From here onward in the text we focus on intrinsic mechanisms that are not related to the specifics of impurity scattering.

The conventional Hall effect (HE) is normally studied with the help of the Boltzmann equation [2, 3]; however, the basics can be qualitatively understood from the classical picture of electron motion in an external magnetic field. In the Hall setting, an external magnetic field is perpendicular to the driving electric field E, as shown in Fig. 2.1.

The Hall voltage proportional to E_H stems from the Lorenz force that deflects electrons in the direction perpendicular to the drift velocity, $\boldsymbol{E}_H = -\mathrm{v} \times \boldsymbol{B}$, where $\boldsymbol{B}$ is the magnetic field. In an isotropic resistive media, the Hall conductivity σ_H^{xy} enters Ohm's law as follows:

$$J_x = \sigma_\Omega E_x + \sigma_H^{xy} E_H,$$
$$J_y = -\sigma_H^{xy} E_x + \sigma_\Omega E_H, \tag{2.1}$$

where

$$\sigma_\Omega = \frac{e^2 n\tau}{m\left(1+\omega_c^2\tau^2\right)}, \quad \sigma_H^{xy} = \omega_c \tau \sigma_\Omega, \quad \omega_c = \frac{eB_z}{m}, \tag{2.2}$$

and n is the carrier density, $\boldsymbol{J}$ is the current density, τ is the average time between scattering events (momentum relaxation time). Onsager reciprocity relations $\sigma^{ij}(\boldsymbol{B}) = -\sigma^{ji}(-\boldsymbol{B})$ imply that the tensor of dissipative (Ohmic) conductivity σ_Ω^{ij} is symmetric and the Hall component σ_H^{ij} is antisymmetric on spatial components ij. So, the Hall component exists only if the time reversal symmetry is broken.

The Hall current, which flows in the y-direction as a response to the x-directed driving electric field, is dissipationless as $\boldsymbol{J} \times \boldsymbol{E} = 0$. In the steady state, the Lorenz force is balanced by the Hall field, $J_y = 0$, and following from (2.1) and (2.2), Hall resistivity is linear in the magnetic field:

$$\rho^{xy} = \frac{E_H}{J_x} = \frac{B_z}{en}. \tag{2.3}$$

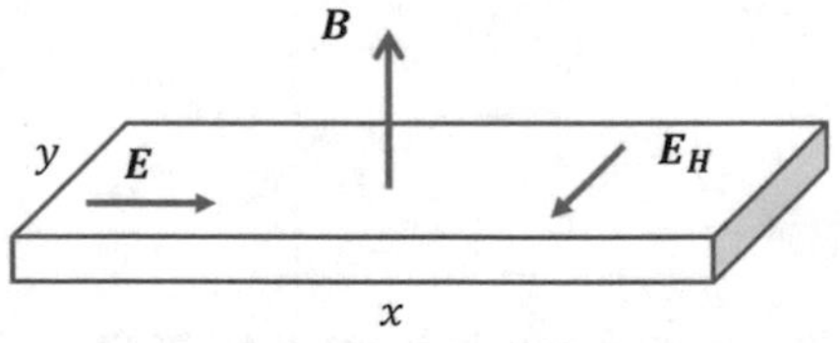

Fig. 2.1 Hall effect geometry

Fig. 2.2 Anomalous Hall resistivity

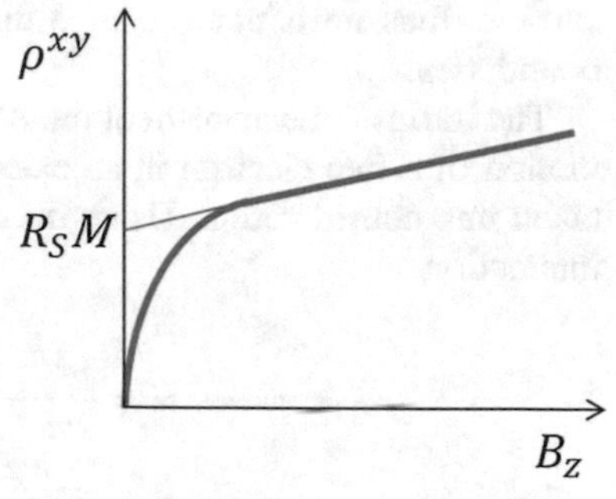

The effect of a magnetic field is two-fold: it breaks the time reversal symmetry and causes transverse orbital motion deflecting carriers with the Lorenz force. Which one is more fundamental for the Hall effect to occur? The answer is not obvious unless one takes into consideration the anomalous Hall effect (AHE) in ferromagnetic metals Fe, Co, Ni, and ferromagnetic spinels [4], where the time reversal symmetry is broken by magnetization, and the large Hall voltage has been observed as directly related, not to the Lorenz force, but rather to magnetization. The anomalous part of Hall resistivity relates to magnetization M aligned along B_z:

$$\rho^{xy} = \frac{B_z}{en} + \rho_S^{xy}, \quad \rho_S^{xy} = R_S M. \tag{2.4}$$

Hall resistivity, shown in Fig. 2.2, starts with the nonlinear part ρ_S^{xy} reflecting the process of coalescence of ferromagnetic domains and then, after magnetic saturation is reached, it becomes proportional to the magnetic field.

Since the anomalous Hall resistivity depends on magnetization, one more observable feature of the effect is the hysteresis loop in ρ^{xy} which follows magnetization when the magnetic field direction is reversed.

In a conventional Hall effect, the "anomalous" electron velocity normal to the driving field appears due to the deflection action of the magnetic field. Quantitatively, the AHE cannot be explained by assuming the conventional Hall setting, where the auxiliary magnetic field is induced by magnetization.

The phenomenology of the effect relates total current and magnetization as follows [5]:

$$\boldsymbol{J} = \sigma_\Omega^{xx} \boldsymbol{E} + \beta \boldsymbol{E} \times \boldsymbol{M}, \tag{2.5}$$

where β is the dissipationless kinetic coefficient that changes its sign under time reversal. The β-term in (2.5) is responsible for the AHE and it originates from the spin-orbit part of the Hamiltonian, $\boldsymbol{p} \cdot \boldsymbol{s} \times \nabla V(\boldsymbol{r})$, so the spin–orbit interaction is essential for AHE to occur. The effect does not exist in the Stoner model of itinerant ferromagnetism as this model does not account for the spin–orbit interaction. Spin–orbit coupling favors electron motion transverse to the driving electric field in opposite directions for opposite spins while magnetization creates majority spin carriers, thus introducing charge disbalance and Hall voltage at opposite sample boundaries.

The intrinsic mechanism of the AHE cannot be understood by assuming classical motion of a free electron in an external electric field even if the spin-orbit term is taken into consideration. Using the effective mass Hamiltonian with the spin–orbit interaction

$$H = \frac{p^2}{2m} + V(\boldsymbol{r}) + A\,\boldsymbol{p} \times \boldsymbol{\nabla} V \cdot \boldsymbol{s}, \tag{2.6}$$

one can write equations of motion as follows:

$$\begin{aligned} \dot{\boldsymbol{r}} &= \partial H/d\boldsymbol{p} = \boldsymbol{p}/m + A\,\boldsymbol{\nabla} V(\boldsymbol{r}) \times \boldsymbol{s}, \\ \dot{\boldsymbol{p}} &= -\frac{\partial H}{\partial \boldsymbol{r}} = -\boldsymbol{\nabla}[V(\boldsymbol{r}) + A\,\boldsymbol{p} \times \boldsymbol{\nabla} V \cdot \boldsymbol{s}] = -\boldsymbol{\nabla} V(\boldsymbol{r}) + Am(\mathbf{v} \times \boldsymbol{s})\Delta V(\boldsymbol{r}), \end{aligned} \tag{2.7}$$

where Δ is the Laplacian, m and $\mathbf{v}$ are the electron mass and velocity, respectively. The second line in (2.7) presents Newton's law and the anomalous force on the right hand side becomes zero if the constant electric field $V(\boldsymbol{r}) = q\boldsymbol{E}\boldsymbol{r}$ is applied as a driving force. So, the effective mass approach, where the only role of a lattice is to renormalize the free electron mass, does not describe AHE as the Hamiltonian (2.6) cannot pick up the correct behavior of electron wave functions in a crystal. The AHE relies on the properties of Bloch amplitudes, periodic factors in electron wave functions which are equal to unity within the free electron approach (2.6). The intrinsic AHE has been explained using the quantum theory of an electron moving in a periodic lattice [6] and will be discussed later on.

If magnetization is equal to zero, the system is time reversal symmetric, the net Hall current is absent, but the transverse spin current is not. Spin current causes spin accumulation at the sample boundaries [7] as illustrated in Fig. 2.3. In phenomenology, the same non-dissipative kinetic coefficient β from (2.5) is responsible for both the AHE and the Spin Hall Effect (SHE) that is transverse to the driving electric field spin accumulation. Spin current J_i^j, defined as the flow of the j-oriented spin polarization along the i-direction, is expressed as

$$J_i^j = \sigma_{ik}^j E_k, \qquad \sigma_{ik}^j \propto \beta \varepsilon_{ijk}, \tag{2.8}$$

where ε_{ijk} is the unit antisymmetric matrix and summation goes over repeating indexes.

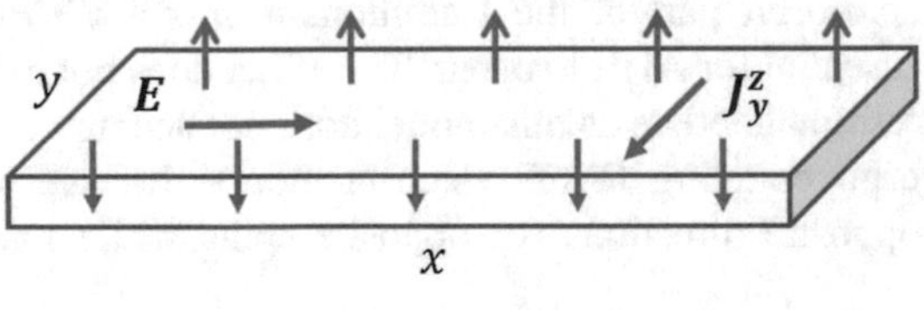

Fig. 2.3 Schematic illustration of the spin Hall effect: z-component spin flow in y-direction under x-oriented electric field

The extrinsic mechanism of the SHE proposed in [7] relies on electron scattering, the process in which, due to spin–orbit coupling, electrons with opposite spins scatter in opposite directions. It appears that, like AHE, SHE also has an intrinsic contribution which does not rely on impurity scattering. So, AHE and SHE share common features: one is the spin–orbit interaction and another, as no magnetic field is needed in their settings, is the non-Lorenz force acting on electrons while they drift under the driving electric field. The nature of this force will be discussed in the next sections.

In various two-dimensional electron systems—inversion layers, quantum wells, semiconductor surfaces, and graphene—all three effects (HE, AHE, and SHE) have their quantum versions QHE, QAHE, and QSHE, respectively. It was recently recognized that those effects and their quantum counterparts are related to the topological properties of Bloch amplitudes. An electron moving in k-space across the BZ under an external field may feel, under certain conditions, the lattice-induced force that is at the origin of the anomalous velocity and the Hall conductivity even in the absence of an external magnetic field. This process can be quantitatively described with the help of the Berry phase, the phase the electron wave function acquires moving slowly along the closed contour in the BZ.

Experimental observation of the SHE has been reported in [8–10]. For a review, see [11].

2.2 Berry Phase

The eigenstate $|n, \boldsymbol{R}(t)\rangle$ of the parameter-dependent Hamiltonian $H(\boldsymbol{R})$ varies along with the slow variation of $\boldsymbol{R}(t)$, the vector in the parameter space. We assume a gap existing between the eigenstate and all other states so the adiabatic flow of $\boldsymbol{R}(t)$ introduces perturbation weak enough to prevent the transition to other states. The picture of the flow assumes $\boldsymbol{R}$-independent boundary conditions, so that all snapshots $|n, \boldsymbol{R}(t)\rangle$ on the path from $\boldsymbol{R}_1$ to $\boldsymbol{R}_2$ belong to the same Hilbert space. The set of snapshots $|n, \boldsymbol{R}(t)\rangle$ along the path forms the space and the Berry phase is closely related to the curvature of that space [12]. In more detail, let us consider the adiabatic evolution in $\boldsymbol{R}$-space. Starting from $\psi_n = |n, \boldsymbol{R}(0)\rangle$, we follow the temporal evolution of the snapshots written with a time-dependent phase factor

$$\psi_n = \exp(i\varphi(t)) \mid n, \boldsymbol{R}(t)\rangle. \tag{2.9}$$

Phase $\varphi(t)$ can be determined if one substitutes the wave function (2.9) into the Schrödinger equation

$$\begin{aligned} i\hbar\dot{\psi}_n &= H(\boldsymbol{R}(t))\psi_n, \\ \langle n, \boldsymbol{R}(t)|n, \boldsymbol{R}(t)\rangle &= 1. \end{aligned} \tag{2.10}$$

Using the equation for an instantaneous eigenvalue $H(\boldsymbol{R}(t)) \mid n, \boldsymbol{R}(t)\rangle = \varepsilon_n(t) \mid n, \boldsymbol{R}(t)\rangle$ and multiplying (2.10) by $\langle n, \boldsymbol{R}(t) \mid \ \exp(-i\varphi_n(t))$, we get the phase as a solution to the equation $\dot{\varphi}_n = i\langle n|\dot{n}\rangle - \varepsilon_n/\hbar$:

$$\varphi_n(T) = i\int_0^T \langle n|\dot{n}\rangle dt + \frac{1}{\hbar}\int_0^T \varepsilon_n dt = \gamma_n(T) + \varphi_D(T). \tag{2.11}$$

The inner product in (2.11) and forward in the text implies spatial integration in the region where the eigenstate is normalized, $\langle n|\, n\rangle = 1$. Two contributions to the phase in (2.11) are defined as the dynamical phase φ_D and the geometrical phase γ_n:

$$\gamma_n(T) = i\int_0^T \langle n|\dot{n}\rangle dt = i\int_{\boldsymbol{R}(0)}^{\boldsymbol{R}(T)} \left\langle n\left|\frac{\partial n}{\partial \boldsymbol{R}}\right.\right\rangle d\boldsymbol{R}. \tag{2.12}$$

For the cyclic evolution, the dynamical phase becomes zero and is not considered below. The integrand in (2.12) is called the Berry connection

$$\boldsymbol{A}(\boldsymbol{R}) = i\left\langle n\left|\frac{\partial n}{\partial \boldsymbol{R}}\right.\right\rangle. \tag{2.13}$$

The name reflects the fact that $\boldsymbol{A}(\boldsymbol{R})$ determines the overlap of eigenstates in two close points in the parametric space, $\boldsymbol{R}$ and $\boldsymbol{R} + \delta\boldsymbol{R}$,

$$\langle n, \boldsymbol{R}|n, \boldsymbol{R} + \boldsymbol{\delta R}\rangle = 1 + \left\langle n\left|\frac{\partial n}{\partial \boldsymbol{R}}\right.\right\rangle \delta\boldsymbol{R} + \cdots = \exp(-i\delta\boldsymbol{R}\cdot\boldsymbol{A}_n(\boldsymbol{R})). \tag{2.14}$$

It is important to note that the evolution described by (2.14) depends on the Berry connection and so is not gauge invariant. This means that the angle between closed eigenvectors depends on the choice of gauge. To compensate the angle, instead of the regular derivative, the covariant derivative $\boldsymbol{D}_{\boldsymbol{R}} = \frac{\partial}{\partial \boldsymbol{R}} + i\boldsymbol{A}(\boldsymbol{R})$ can be used in (2.14). This would imply $\langle n|\boldsymbol{D}_{\boldsymbol{R}} n\rangle$ instead of $\langle n|\frac{\partial n}{\partial \boldsymbol{R}}\rangle$, thus providing for the gauge invariant evolution with a scalar product (2.14) equal to unity. So, the Berry vector potential $\boldsymbol{A}(\boldsymbol{R})$ naturally emerges in the parallel transport manifesting a non-trivial topology of the Hilbert space to which the wave function $|n, \boldsymbol{R}(t)\rangle$ belongs.

Under local phase transformation with smooth function $g(\boldsymbol{R})$ ($U(1)$ gauge transformation)

$$|n, \boldsymbol{R}(t)\rangle \to |n, \boldsymbol{R}(t)\rangle \exp(ig(\boldsymbol{R})), \tag{2.15}$$

the Berry connection (2.13) and the γ_n-phase (2.12) are transformed as

$$A_n(\boldsymbol{R}) \to A_n(\boldsymbol{R}) - \frac{\partial g(\boldsymbol{R})}{\partial \boldsymbol{R}}, \quad \gamma_n \to \gamma_n + g(\boldsymbol{R}(T)) - g(\boldsymbol{R}(0)). \tag{2.16}$$

So, the local phase transformation in the parameter space (2.15) applied to the wave function results in the gauge transformation of the Berry connection (2.16) exactly as the local phase shift in the electron wave function in real space is equivalent to the gauge transformation of the electromagnetic vector potential. Because of this formal analogy, the Berry connection is also called the Berry vector potential.

The ambiguity of the Berry connection relates to the choice of the gauge $g(\boldsymbol{R})$. Then the γ_n-phase (2.16) can become zero by the appropriate choice of gauge. There is one exception that makes the geometrical phase gauge invariant: for a cyclic evolution, the integration in (2.12) goes over the closed path C and the phase ambiguity disappears:

$$\boldsymbol{R}(T) = \boldsymbol{R}(0), \quad g(\boldsymbol{R}(T)) - g(\boldsymbol{R}(0)) = 2\pi n, \quad n = 0,1,2\ldots \tag{2.17}$$

The gauge invariant cyclic γ_n-phase is known as the Berry phase:

$$\gamma_n(C) = \oint_C \boldsymbol{A}_n(\boldsymbol{R}) \cdot d\boldsymbol{R} \quad mod\ 2\pi \tag{2.18}$$

The gauge field tensor corresponding to the Berry vector potential is called the Berry curvature (Berry field)—analogous to a magnetic field in real space:

$$F_{n\alpha\beta} = \frac{\partial A_{n,\beta}}{\partial R_\alpha} - \frac{\partial A_{n,\alpha}}{\partial R_\beta} = i\left(\left\langle \frac{\partial n}{\partial R_\alpha}\middle|\frac{\partial n}{\partial R_\beta}\right\rangle - \left\langle \frac{\partial n}{\partial R_\beta}\middle|\frac{\partial n}{\partial R_\alpha}\right\rangle\right) = -2\,\mathrm{Im}\left\langle \frac{\partial n}{\partial R_\alpha}\middle|\frac{\partial n}{\partial R_\beta}\right\rangle. \tag{2.19}$$

As expression (2.19) contains the difference of derivatives, the ambiguity in A_n is cancelled out and the Berry field remains gauge invariant. The Berry field can be recast into an expression that does not contain derivatives of eigenvectors. As instantaneous eigenvectors of the Hamiltonian form a complete and orthonormalized set, we use the relation $\sum_m | m\rangle\langle m | = 1$ in order to present (2.19) in the form

$$\left\langle \frac{\partial n}{\partial R_\alpha}\middle|\frac{\partial n}{\partial R_\beta}\right\rangle = \sum_m \left\langle \frac{\partial n}{\partial R_\alpha}\middle| m\right\rangle \left\langle m \middle|\frac{\partial n}{\partial R_\beta}\right\rangle. \tag{2.20}$$

Calculating derivatives $\frac{\partial}{\partial R_\alpha}\langle m|H|n\rangle$ and $\frac{\partial}{\partial R_\alpha}\langle m|n\rangle$ one gets

$$\left\langle \frac{\partial n}{\partial R_\alpha}\middle| H \middle| n\right\rangle + \left\langle m \middle| \frac{\partial H}{\partial R_\alpha} \middle| n\right\rangle + \left\langle m \middle| H \middle| \frac{\partial n}{\partial R_\alpha}\right\rangle = \frac{\partial \varepsilon_n}{\partial R_\alpha}\delta_{mn},$$

$$\frac{\partial}{\partial R_\alpha}\langle m|n\rangle = \left\langle \frac{\partial m}{\partial R_\alpha}\middle| n\right\rangle + \left\langle m \middle| \frac{\partial n}{\partial R_\alpha}\right\rangle = 0. \tag{2.21}$$

If $n \neq m$, (2.21) gives

$$\left\langle n \middle| \frac{\partial H}{\partial R_\alpha} \middle| m\right\rangle = (\varepsilon_n - \varepsilon_m)\left\langle \frac{\partial n}{\partial R_\alpha}\middle| m\right\rangle, \quad n \neq m. \tag{2.22}$$

Using (2.22), we exclude $\left\langle \frac{\partial n}{\partial R_\alpha}\middle| m\right\rangle$ from (2.20) and come to the Berry field given as

$$F_{n,\alpha\beta} = -2\,\mathrm{Im}\sum_{m\neq n}\left\langle n \middle| \frac{\partial H}{\partial R_\alpha} \middle| m\right\rangle\left\langle m \middle| \frac{\partial H}{\partial R_\beta} \middle| n\right\rangle/(\varepsilon_n - \varepsilon_m)^2. \tag{2.23}$$

The Berry field is singular in the parameter space region where the spectrum is degenerate. The sum of Berry fields over all bands consists of complex conjugate contributions, so the imaginary part in (2.23) vanishes: $\sum_n F_{n,\alpha\beta} = 0$. A restricted number of eigenstates out of the whole set might feel nonzero Berry field. This fact reflects the need to allow the existence of the "outside world" for the eigenstate whose adiabatic evolution we follow.

If the parametric space is of three or fewer dimensions, the Berry field (2.19) can be expressed as a vector cross-product,

$$\boldsymbol{F}_n = \boldsymbol{\nabla_R} \times \boldsymbol{A}_n(\boldsymbol{R}) = \boldsymbol{i}\langle \boldsymbol{\nabla_R n}| \times |\boldsymbol{\nabla_R n}\rangle,$$

$$F_n^\alpha(\boldsymbol{R}) = \varepsilon_{\alpha\beta\gamma}\frac{\partial}{\partial R_\beta}A_{n,\gamma}(\boldsymbol{R}). \tag{2.24}$$

Some general properties of the Berry field can be understood by exploring the spin moment in a magnetic field as an example.

2.3 Spin in a Magnetic Field: Magnetic Monopole

Let us follow the evolution of a spin-1/2 wave function when the direction of the magnetic field $\boldsymbol{B}$ rotates along the circle as shown in Fig. 2.4. Slow rotation preserves the spin alignment along the magnetic field at each point on the evolution path. Our two-dimensional parameter space consists of angles (ϕ, θ) that determine the direction of the magnetic field.

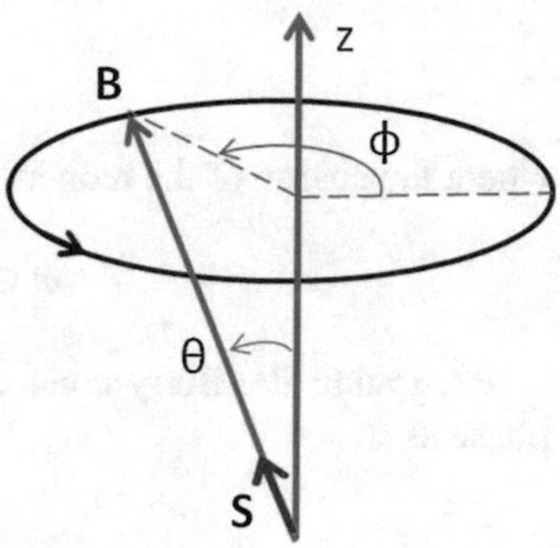

Fig. 2.4 Cyclic spin transport by magnetic field rotation

A spin-1/2 particle in a magnetic field is described by the Hamiltonian $H = -\boldsymbol{B}\cdot\boldsymbol{\sigma}$, where $\boldsymbol{B}$ is the magnetic field in energy units, and $\boldsymbol{\sigma}$ is the vector comprising Pauli matrices. As long as the magnetic field is pointed at the upper hemisphere, $0 < \theta \leq \pi/2$, the ground state eigenvector of a (ϕ, θ)-oriented spin is expressed as

$$|n\rangle = \begin{pmatrix} \cos\frac{\theta}{2} \\ e^{i\phi}\sin\frac{\theta}{2} \end{pmatrix}. \tag{2.25}$$

The Berry connection on the circle is calculated below:

$$dl = B\sin\theta\, d\phi,$$

$$\left|\frac{\partial n}{\partial l}\right\rangle = \left|\frac{\partial n}{\partial \phi}\frac{\partial \phi}{\partial l}\right\rangle = \frac{1}{B\sin\theta}\left|\frac{\partial n}{\partial \phi}\right\rangle,$$

$$A_\phi = i\left\langle n\left|\frac{\partial n}{\partial l}\right.\right\rangle = \frac{1-\cos\theta}{2B\sin\theta} \tag{2.26}$$

The Berry field, expressed as $\boldsymbol{F} = \boldsymbol{\nabla_B} \times \boldsymbol{A}$, can be calculated by representing the ***curl*** operation in spherical coordinates and taking into account that on the evolution path $B = |\boldsymbol{B}|$ and $A_\theta = A_B = 0$. The result has the form of the field of a magnetic monopole:

$$\boldsymbol{F} = \boldsymbol{\nabla_B} \times \boldsymbol{A} = \left(\frac{1}{B\sin\theta}\frac{\partial}{\partial\theta}A_\phi\sin\theta\right)\boldsymbol{b} = \frac{\boldsymbol{b}}{2B^2}, \tag{2.27}$$

where $\boldsymbol{b}$ is the unit vector along $\boldsymbol{B}$, $\boldsymbol{b} = \boldsymbol{B}/B$. For spin-down eigenstate, the sign in (2.27) is reversed.

Generalization to higher spin-momentum S results in the expression for the Berry field

$$F = m\frac{\boldsymbol{b}}{B^2}, \tag{2.28}$$

where the charge of the monopole depends on the state under consideration:

$$m \in \{-S, -S+1, \ldots, S\}.$$

Integrating the Berry connection along the path in (2.18), one gets the Berry phase as

$$\gamma = \oint A_\phi dl = \frac{1}{2}(1 - \cos\theta)\int_0^{2\pi} d\phi = \pi(1 - \cos\theta) = \frac{1}{2}\Omega_l \quad mod\ 2\pi, \tag{2.29}$$

where Ω_l is the solid angle subtended by the closed contour l illustrated in Fig. 2.4. The same Berry phase can be obtained with the help of the Stokes theorem, integrating the Berry field over the spherical surface enclosed by the path l:

$$\gamma = \int \boldsymbol{F}\, d\boldsymbol{S} = \frac{1}{2}\int_0^{\theta} \sin\theta\, d\theta \int_0^{2\pi} d\phi = \pi(1 - \cos\theta). \tag{2.30}$$

It follows from (2.29) that transporting the spin along the path l at $\theta = \pi/2$ we rotate it through 2π and acquire the Berry phase of π that is equivalent to the factor -1 in the spinor (2.25):

$$|n\rangle \to \exp(i\gamma)|n\rangle = -|n\rangle. \tag{2.31}$$

As for gauge invariance, a change in phase in the spinor (2.25) would change the phase-dependent expression for the Berry connection A_ϕ while the Berry curvature F is gauge invariant and formally looks like the field of a magnetic monopole as it diverges, manifesting degeneracy of the spectrum at $B \to 0$ (see [13] for details).

A non-trivial Berry phase of π appears in (2.30) if the path l encircles the degeneracy point or, equivalently, lies in non-simply-connected 2D-domain in the parameter space. Integration of the Berry field over the whole sphere ($\theta = \pi$ in (2.30)) gives 2π. More generally, the integral (2.30) over a closed surface without a border must be quantized as

$$\int \boldsymbol{F}\, d\boldsymbol{S} = 2\pi C_1, \tag{2.32}$$

where $C_1 \in Z$ is the topological invariant of the system called the first Chern number [14] (see [15] for details). The monopole is a source (drain) of the Berry field, and C_1, counts the net number of monopoles inside the closed surface, providing pairs of monopoles of opposite signs (sources and drains) annihilate. In the particular example discussed here, the Chern number is equal to ± 1 for spin up and down, respectively.

It should be noted that the monopole-type Berry field follows from the shape of the spinor (2.25) which cannot be well defined in both upper and lower hemispheres, i.e., at $\theta = 0$ and at $\theta = \pi$. At $\theta = 0$ the ground state spinor $|n\rangle$ is well defined and presents the +1 eigenstate of σ_z, $\begin{pmatrix} 1 \\ 0 \end{pmatrix}$, whereas at $\theta = \pi$ it is equal to $\begin{pmatrix} 0 \\ \exp(i\phi) \end{pmatrix}$ and can be converted into the correct -1 eigenstate $\begin{pmatrix} 0 \\ 1 \end{pmatrix}$ only after multiplication by the phase factor of $\exp(-i\phi)$. So, the eigenstate $|n\rangle$ cannot be defined globally, on a whole sphere. Similar considerations apply to the corrected spinor $\exp(-i\phi)|n\rangle$, which is well defined at $\theta = \pi$ and ill-defined at $\theta = 0$. It manifests the necessity of the transition function $\exp(-i\phi)$ in order to overcome an obstruction to defining eigenstates globally. The source or drain that appears in the integral (2.33) is a consequence of this obstruction: if in a quantum-mechanical system there is no such obstruction, the system is topologically trivial and the integral constructed like that in (2.33) becomes zero.

The Chern number and 2π-quantization appear also in two-dimensional solids, where the Hamiltonian has no special symmetry except of spatial translation invariance. This problem will be discussed later in relation to the QHE. Some symmetry operations may call for another type of quantization. An example is the Hamiltonian invariant to an anti-unitary operation of complex conjugate K, $K^2 = 1$ [16]. In the system of spinless particles, the K-transformation is the time reversal operation that leaves the eigenstate unchanged. In such a system, the Berry connection (2.13) is imaginary, $\boldsymbol{A}_{Kn} = \boldsymbol{A}_n^* = -\boldsymbol{A}_n$. As $|n\rangle$ and $K|n\rangle$ describe the same state and the Berry phase is defined as *mod* 2π, it follows that

$$\gamma_n(C) = -\gamma_{Kn}(C) = -\gamma_n(C) \quad mod\ 2\pi. \tag{2.33}$$

Equation (2.33) can be satisfied if the Berry phase is π-quantized:

$$\gamma_n(C) = \begin{cases} 0 \\ \pi \end{cases} \quad mod\ 2\pi. \tag{2.34}$$

The same type of quantization has been found in Zak's phase, where it is caused by spatial inversion symmetry. Zak's phase (see the next section) is the Berry phase in a crystalline solid where the integration path transverses the Brillouin zone and can be considered as closed due to the periodic gauge [17, 18].

The Berry phase and the Berry field (curvature) are observables and manifest themselves in many fields of solid state and molecular physics (see [17, 19, 20] for review). A parameter-dependent one-electron Hamiltonian in solids presents the scene where the Berry phase plays its role in the most natural way.

2.4 Crystalline Solids: Berry and Zak Phases

In the periodic lattice potential, the electron Hamiltonian is invariant to lattice translations $\boldsymbol{R}$, so the wave function is an eigenfunction of the translation operator

$$\psi(\boldsymbol{r}+\boldsymbol{R}) = \psi(\boldsymbol{r})\exp(i\boldsymbol{k}\boldsymbol{R}), \tag{2.35}$$

where $\exp(i\boldsymbol{k}\boldsymbol{R})$ is the eigenvalue of the translation operator and $\boldsymbol{k}$ is the electron wave vector. Periodicity (2.35) and Born-von Karman cyclic boundary conditions dictate the shape of the wave function and the structure of the $\boldsymbol{k}$-space. The eigenfunctions of a one-electron Hamiltonian that includes the periodic potential are the Bloch functions $\psi_{n\boldsymbol{k}}(\boldsymbol{r}) = u_{n\boldsymbol{k}}(\boldsymbol{r})\exp(i\boldsymbol{k}\boldsymbol{R})$, where $u_{n\boldsymbol{k}}(\boldsymbol{r})$ are the unit cell periodic Bloch amplitudes, n labels the energy band, and the wave vector is defined modulo $\boldsymbol{G}$, the reciprocal lattice vector. So, the base space of our parametric Hamiltonian is the first Brillouin zone. The one-electron Hamiltonian can be written in k-representation $H_{\boldsymbol{k}} = \exp(-i\boldsymbol{k}\boldsymbol{r})H\exp(i\boldsymbol{k}\boldsymbol{r})$ (see (1.1) as an example). In this representation, the Hamiltonian depends on parameters $\boldsymbol{k}$ and the eigenstates are the Bloch amplitudes $u_{n\boldsymbol{k}}(\boldsymbol{r})$ which satisfy $\boldsymbol{k}$-independent boundary conditions. This allows us to use the Berry phase concept in order to describe adiabatic evolution in $\boldsymbol{k}$ space using (2.12):

$$\gamma_n(T) = i\int_{\boldsymbol{k}(0)}^{\boldsymbol{k}(T)} \left\langle u_{n\boldsymbol{k}} \middle| \frac{\partial u_{n\boldsymbol{k}}}{\partial \boldsymbol{k}} \right\rangle d\boldsymbol{k}, \tag{2.36}$$

Phase $\gamma_n(T)$ becomes gauge invariant if the integration path closes inside the BZ, in which case this phase is nothing more than what was discussed in Sect. 2.3. However, in solids there is the option to close the path, integrating over the whole BZ. As functions $\psi_{n\boldsymbol{k}}(\boldsymbol{r})$ and $\psi_{n\boldsymbol{k}+\boldsymbol{G}}(\boldsymbol{r})$ describe the same electron state, they can be chosen to be equal (periodic gauge), resulting in Bloch amplitudes at $\boldsymbol{k}$ and $\boldsymbol{k}+\boldsymbol{G}$ related by $\boldsymbol{k}$-independent unitary gauge transformation $u_{n\boldsymbol{k}} = \exp(i\boldsymbol{G}\boldsymbol{r})u_{n\boldsymbol{k}+\boldsymbol{G}}$. The integral over the open path is gauge independent if the wave functions at the initial and final points are related by a unitary transformation as it preserves the phase coherence between them [18, 21]. This phase was introduced for 1D solids in [17] and is called the Zak phase. In two dimensions, the Zak phase is expressed as an integral over the BZ:

$$\gamma_n = i\int_{\mathrm{BZ}} \langle u_{n\boldsymbol{k}} | \nabla_{\boldsymbol{k}} u_{n\boldsymbol{k}} \rangle d\boldsymbol{k}. \tag{2.37}$$

Another way to look at the problem is to use the fact that from the topology standpoint, the 2D BZ is a 2D torus, and then the integration over k_x and k_y runs along the closed paths as points $\boldsymbol{k}$ and $\boldsymbol{k}+\boldsymbol{G}$ are physically equivalent. The torus is a compact manifold and the Berry phase, being the integral of Berry curvature over the closed surface, should be quantized in accordance with the Chern theorem (see (2.32)):

$$\gamma_n = \int_{\mathrm{BZ}} \boldsymbol{F}_n d\boldsymbol{S} = 2\pi C_1 \quad mod\, 2\pi,$$
$$\boldsymbol{F}_n = \nabla_{\boldsymbol{k}} \times \boldsymbol{A}_n(\boldsymbol{k}). \tag{2.38}$$

As we follow adiabatic transport, we assume there is an energy gap in the electron spectrum and the whole bundle of n eigenfunctions u_{nk} is divided into two sub-bundles: valence bands filled with electrons and empty conduction bands. If γ_n in a valence band is nonzero we are dealing with a Chern insulator, otherwise we are dealing with a conventional insulator. One needs at least 2D parameter space in order to arrange integration over a surface and get quantization in (2.38), so the Chern insulator exists only in a 2D crystal, where the parameter space, the Brillouin zone, is a 2D torus. A most interesting example of a Chern insulator is the quantum Hall system: 2D-electrons in the plane perpendicular to the magnetic field.

As follows from the semiclassical analyses of electron motion in band n under applied fields [19, 20, 22, 23], the Berry field $\boldsymbol{F}_n(\boldsymbol{k})$ enters the equations of motion as follows:

$$\dot{\boldsymbol{r}} = \frac{\partial \varepsilon}{\hbar \partial \boldsymbol{k}_0} - \dot{\boldsymbol{k}}_0 \times \boldsymbol{F}_n(\boldsymbol{k}_0),$$
$$\hbar \dot{\boldsymbol{k}}_0 = -e\boldsymbol{E} - e\dot{\boldsymbol{r}} \times \boldsymbol{B}, \tag{2.39}$$

where ε and $\boldsymbol{k_0}$ are the energy and wave vector of the electron wave packet, respectively. Taking the anomalous part of (2.39) $\dot{\boldsymbol{r}} \sim \dot{\boldsymbol{k}} \times \boldsymbol{F}$ and interchanging $\boldsymbol{k} \leftrightarrow \boldsymbol{r}$ and $\boldsymbol{F} \leftrightarrow \boldsymbol{B}$ (magnetic and Berry fields), one gets $\dot{\boldsymbol{k}} \sim \dot{\boldsymbol{r}} \times \boldsymbol{B}$, equivalent to the equation of motion under a Lorenz force (second line in (2.39)). Formally, interchanging $\boldsymbol{k} \leftrightarrow \boldsymbol{r}$ means a Fourier transformation from $\boldsymbol{k}$-space to real space, so, the Berry field acts in $\boldsymbol{k}$-space like a magnetic field in real space. The duality between real and the momentum space implies that $\boldsymbol{B}$ and $\boldsymbol{F}$ are independent and one may exist without the other.

2.5 Hall Conductivity and Berry Phase

Within the linear response approach, static Hall conductivity is given by the Kubo formula which can be presented in the form [24]:

$$\sigma_{xy} = \frac{2e^2\hbar}{(2\pi)^3} \sum_{\substack{m,n \\ m \neq n}} \int_{\mathrm{BZ}} dk f_n(\boldsymbol{k}) \frac{\mathrm{Im}[\langle \psi_{nk} \mid \mathrm{v}_x \mid \psi_{mk} \rangle \langle \psi_{mk} \mid \mathrm{v}_y \mid \psi_{nk} \rangle]}{(\varepsilon_{nk} - \varepsilon_{mk})^2}, \tag{2.40}$$

where $f_n(\boldsymbol{k})$ is the Fermi filling factor, $\psi_{n\boldsymbol{k}}$ are the Bloch functions, and the eigenfunction of the Hamiltonian H, $\mathbf{v} = \partial H/\partial \boldsymbol{p}$ is the velocity operator. Substituting Bloch functions in (2.40) and taking into account relations $H_{\boldsymbol{k}} = e^{-i\boldsymbol{kr}} H e^{i\boldsymbol{kr}}$, $\nabla_{\boldsymbol{k}} H_{\boldsymbol{k}} = \hbar e^{-i\boldsymbol{kr}} \mathbf{v} e^{i\boldsymbol{kr}}$, one obtains the Hall conductivity expressed through the Hamiltonian $H_{\boldsymbol{k}}$ defined in the parameter space-Brillouin zone:

$$\begin{aligned}\sigma_{xy} &= \frac{2e^2}{\hbar(2\pi)^3} \sum_{\substack{m,n \\ m \neq n}} \int_{\mathrm{BZ}} d\boldsymbol{k} [f_n(\boldsymbol{k}) - f_m(\boldsymbol{k})] \frac{\mathrm{Im}\left[\langle u_{n\boldsymbol{k}} | \nabla_{k_x} H_{\boldsymbol{k}} | u_{m\boldsymbol{k}} \rangle \langle u_{m\boldsymbol{k}} | \nabla_{k_y} H_{\boldsymbol{k}} | u_{n\boldsymbol{k}} \rangle\right]}{(\varepsilon_{n\boldsymbol{k}} - \varepsilon_{m\boldsymbol{k}})^2} \\ &= -\frac{e^2}{\hbar(2\pi)^3} \sum_n \int_{\mathrm{BZ}} d\boldsymbol{k}\, f_n(\boldsymbol{k})\, \boldsymbol{F}_{nz},\end{aligned} \tag{2.41}$$

where n-summation goes over filled bands and

$$\boldsymbol{F}_{nz}(\boldsymbol{k}) = -2\ \mathrm{Im} \sum_{\substack{m \\ m \neq n}} \frac{\langle u_{n\boldsymbol{k}} | \nabla_{k_x} H_{\boldsymbol{k}} | u_{m\boldsymbol{k}} \rangle \langle u_{m\boldsymbol{k}} | \nabla_{k_y} H_{\boldsymbol{k}} | u_{n\boldsymbol{k}} \rangle}{(\varepsilon_{n\boldsymbol{k}} - \varepsilon_{m\boldsymbol{k}})^2}. \tag{2.42}$$

The structure of the expression for $\boldsymbol{F}_n(\boldsymbol{k})$ is identical to that of the Berry field in (2.23). Equation (2.41) relates the Hall conductivity to the total Berry field that comprises contributions from all filled bands.

To follow how the crystal symmetry affects $\boldsymbol{F}_n(\boldsymbol{k})$, it is convenient to look at the Berry connection and the Berry field both expressed through Bloch amplitudes (see (2.13) and (2.24)):

$$\begin{aligned}\boldsymbol{A}_n(\boldsymbol{k}) &= i\langle u_{n\boldsymbol{k}} | \nabla_{\boldsymbol{k}} u_{n\boldsymbol{k}} \rangle, \\ \boldsymbol{F}_n(\boldsymbol{k}) &= i\langle \nabla_{\boldsymbol{k}} u_{n\boldsymbol{k}} | \times | \nabla_{\boldsymbol{k}} u_{n\boldsymbol{k}} \rangle.\end{aligned} \tag{2.43}$$

First, we consider the spinless particles in a time reversal (TR) and spatial inversion (SI) symmetric solid with eigenstates transforming as $u_{n\boldsymbol{k}}(\boldsymbol{r}) = u_{n,-\boldsymbol{k}}(-\boldsymbol{r})$ and $u_{n\boldsymbol{k}}(\boldsymbol{r}) = u^*_{n,-\boldsymbol{k}}(\boldsymbol{r})$ under SI and TR operations, respectively. These symmetry constrains translate into relations for the Berry field as $\boldsymbol{F}_n(\boldsymbol{k}) = \boldsymbol{F}_n(-\boldsymbol{k})$ under SI and $\boldsymbol{F}_n(\boldsymbol{k}) = -\boldsymbol{F}_n(-\boldsymbol{k})$ under TR-transformation. This means that as long as we deal with non-degenerate energy bands with SI and TR-symmetry, the Berry field and then the Hall conductivity is identically zero. Berry field-related intrinsic Hall effects occur in systems where the TR-symmetry is broken by the magnetic field in the HE or by magnetization in the AHE.

To analyze AHE, we need to take spin into consideration. A TR-symmetric system is Kramers degenerate. The symmetry operations act on eigenstates as $u_{n\boldsymbol{k}\uparrow}(\boldsymbol{r}) = u_{n,-\boldsymbol{k}\uparrow}(-\boldsymbol{r})$ (SI) and $u_{n\boldsymbol{k}\uparrow}(\boldsymbol{r}) = u^*_{n,-\boldsymbol{k}\downarrow}(\boldsymbol{r})$ (TR), so the Berry connection and Berry field defined previously for non-degenerate states should be re-defined to account for the Kramers degeneracy. Re-defined quantities from (2.43) become the non-Abelian Berry connection (vector-valued 2×2 matrix) $A_{nij}(\boldsymbol{k}) = i\langle u_{ni}(\boldsymbol{k})|\nabla_k u_{nj}(\boldsymbol{k})\rangle$, $i, j = \uparrow, \downarrow$. The corresponding non-Abelian Berry field contains an additional commutator term which provides for the gauge invariance [25]:

$$\boldsymbol{F}_{nij}(\boldsymbol{k}) = i\langle \nabla_{\boldsymbol{k}} u_{ni}(\boldsymbol{k})| \times |\nabla_{\boldsymbol{k}} u_{nj}(\boldsymbol{k})\rangle - i \sum_{s=\uparrow\downarrow} \langle \nabla_{\boldsymbol{k}} u_{ni}(\boldsymbol{k})|\nabla_{\boldsymbol{k}} u_{ns}(\boldsymbol{k})\rangle \times \langle \nabla_{\boldsymbol{k}} u_{ns}(\boldsymbol{k})|\nabla_{\boldsymbol{k}} u_{nj}(\boldsymbol{k})\rangle. \quad (2.44)$$

Gauge transformation in two-fold degenerate spin subspace is equivalent to $SU(2)$ spin rotation. Since it is a 2×2 matrix, the Berry field is proportional to the electron spin matrix and following (2.39) generates an anomalous velocity of opposite directions for spin-up and spin-down electrons. The gauge invariant quantity $\boldsymbol{Tr}[\boldsymbol{F}_{ij}]$ replaces $\boldsymbol{F}$ in (2.41) and becomes zero in the TR-invariant system as $\boldsymbol{F}_{\uparrow\uparrow}(\boldsymbol{k}) = -\boldsymbol{F}_{\downarrow\downarrow}(\boldsymbol{k})$, so the Hall conductivity becomes zero, not necessarily because the Berry field identically vanishes, but because anomalous velocities of opposite signs cancel each other and turn the net transverse current to zero. If the cancellation is broken by magnetization, the unequal number of spin-up and spin-down carriers results in the net Hall current called the intrinsic AHE.

In order to get the nonzero Berry fields in spin-resolved channels, the band spectrum has to satisfy the conditions discussed in [26], namely, the band spectrum is accidentally degenerate and the spin–orbit interaction along with magnetization lifts the degeneracy, generating the Berry field of opposite sign for spin-up and spin-down branches of the electron spectrum. One can follow these conditions in ferromagnetic metals, where before the spin–orbit interaction is taken into account, the magnetic bands are spin-split by the exchange interaction so the spin subbands are degenerate at some points of BZ, as shown in Fig. 2.5.

Since the degeneracy is accidental, i.e., not required by the crystal symmetry, it is lifted by the spin–orbit interaction, as illustrated in Fig. 2.6. The spin–orbit interaction repels levels at degeneracy points, creating k-regions with strong spin-momentum dependence. These regions generate the Berry field and result in AHE in metals.

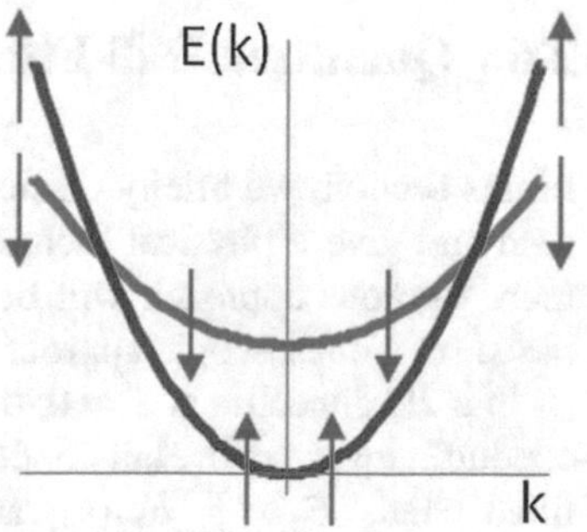

Fig. 2.5 Accidental deneracy of magnetic subbands. Arrows indicate electron spin direction

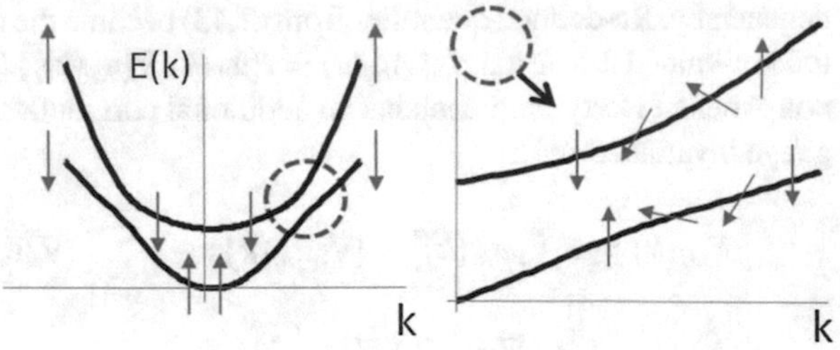

Fig. 2.6 The degeneracy region is the spin-orbit-induced source of the Berry field

If the magnetization is zero, the TR-symmetry keeps the charge Hall current at zero and provides the setting for SHE. Spin Hall current is finite in a TR-symmetric system, and it is determined by the Berry field-like Kubo expression for spin Hall conductivity [27, 28]. Using the notation in (2.8), we present the spin Hall conductivity in the setup shown in Fig. 2.3. Spin current here is the flow in the *y*-direction of the spin *z*-component under the *x*-oriented driving electric field:

$$\sigma_{xy}^{z} = -\frac{2e\hbar}{(2\pi)^3}\sum_n \int_{\mathrm{BZ}} d\boldsymbol{k}\, f_n(\boldsymbol{k}) \sum_{m\neq n} \frac{\mathrm{Im}\left[\langle\psi_{nk}|J_x^z|\psi_{mk}\rangle\langle\psi_{mk}|\mathrm{v}_y|\psi_{nk}\rangle\right]}{(\varepsilon_{nk}-\varepsilon_{mk})^2}, \quad (2.45)$$

where $\mathbf{v}$ and $J_x^z = \frac{1}{4}(\sigma_z \mathrm{v}_x + \mathrm{v}_x \sigma_z)$ are the 2×2 matrix velocity and spin current operators, respectively, and ψ_{nk} is the 2-spinor Bloch eigenstate in band n.

As mentioned above, in TR-symmetric solids the Hall conductivity is absent. This follows from the Berry curvature (2.42). TR reverses velocities and acts as a complex conjugation on wave functions changing the sign of the of Berry curvature, $\boldsymbol{F}(\boldsymbol{k}) = -\boldsymbol{F}(-\boldsymbol{k})$, which being integrated in BZ, becomes zero. Spin Hall conductivity (2.45), on the contrary, contains an additional sign change due to the spin matrix in J_x^z, so it does not change sign under the time reversal that results in finite SHE in TR-symmetric solids.

In 2D-materials, the response (2.41) can be quantized and can be the background for the explanation of the family of quantum effects briefly discussed in the following sections. More details will be given in Chap. 3 where we directly calculate the Berry field in relation to various Hall effects.

2.6 Quantized Hall Effects

In this section, we briefly discuss quantized Hall effects as they relate to the Berry field and give a physical picture of edge currents without detailed calculations. A more rigorous approach will be given in Chap. 3, where the QAHE is calculated based on a microscopic approach.

In a 2D dielectric at $T = 0$, the Fermi level is in the energy gap and the diagonal conductivity $\sigma_{xx} = 0$. Hall conductivity (2.41) contains the total Berry field from all filled bands $\boldsymbol{F}_z = \sum_n \boldsymbol{F}_{nz}(\boldsymbol{k})$, and the integration goes over the two-dimensional closed surface-BZ torus. The Chern theorem (2.38) makes the Hall conductivity quantized as:

$$\sigma_{xy} = -\frac{e^2}{\hbar(2\pi)^2}\int_{\mathrm{BZ}} dk\, \boldsymbol{F}_z = -\frac{e^2}{h}C_1. \tag{2.46}$$

The concept of a Chern insulator allows the treatment of various intrinsic quantum Hall effects within a common approach.

2.6.1 Integer Quantum Hall Effect

The quantum Hall setup shown in Fig. 2.1 implies that the $2D$-crystal with TR-symmetry, violated by an external magnetic field, turns into a Chern insulator, where $C_1 \neq 0$ is the topological invariant. As C_1 is quantized as an integer, the conductivity (2.46) cannot change continuously with the variation of the eigenstate thus making the system robust against scattering and many-body interactions. Experimentally, the integer QHE (IQHE) has been reported in [29].

The nonzero integral in (2.46) has been discussed with regard to an obstruction to define the eigenstate globally on the whole surface of integration (see text after (2.33)). Here we deal with integration over the torus in k-space which is equivalent to a rectangle with periodic boundary conditions. We perform explicit integration in (2.46) assuming the lattice parameter $a = 1$:

$$\begin{aligned} 2\pi C_1 &= \int_{-\pi}^{\pi} dk_x \int_{-\pi}^{\pi} dk_y \left(\frac{\partial A_y}{\partial k_x} - \frac{\partial A_x}{\partial k_y}\right) \\ &= \int_{-\pi}^{\pi} dk_y \left[A_y(\pi, k_y) - A_y(-\pi, k_y) - \frac{\partial}{k_y}\int_{-\pi}^{\pi} dk_x A_x(k_x, k_y)\right]. \end{aligned} \tag{2.47}$$

The periodic gauge in the k_x direction does not depend on momentum (see discussion following (2.36)), so the gauge potential on opposite boundaries of BZ satisfies the relation $A_y(\pi, k_y) = A_y(-\pi, k_y)$, reducing (2.47) to the form:

$$\begin{aligned} 2\pi C_1 &= -\int_{-\pi}^{\pi} dk_y \frac{\partial}{k_y}\int_{-\pi}^{\pi} dk_x A_x(k_x, k_y) = -\int_{-\pi}^{\pi} d\gamma(k_y), \\ \gamma(k_y) &= \int_{-\pi}^{\pi} dk_x A_x(k_x, k_y), \end{aligned} \tag{2.48}$$

where $\gamma(k_y)$ is the Berry phase acquired on the k_x closed loop. Since the Berry phase relates Bloch amplitudes at BZ boundaries, where amplitudes are physically equivalent, $\gamma(k_y)$ is quantized mod 2π. The integral in (2.48) is equal to the full angle comprising 2π k_x-rotations when the eigenstate is moving across BZ in the k_y

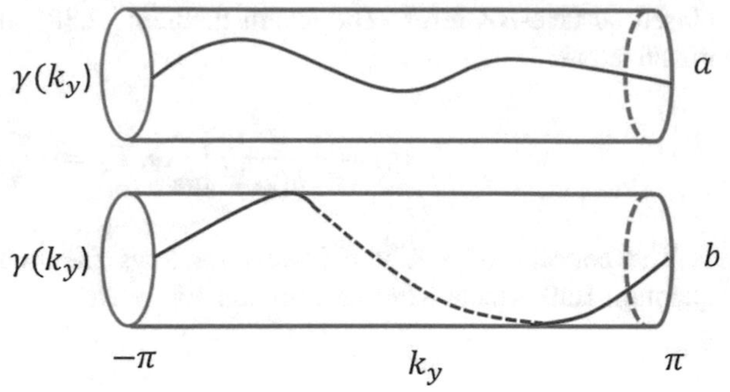

Fig. 2.7 Chern invariant as a winding number. (**a**) normal dielectric, $C_1 = 0$; (**b**) twisted boundary condition, $C_1 = 1$

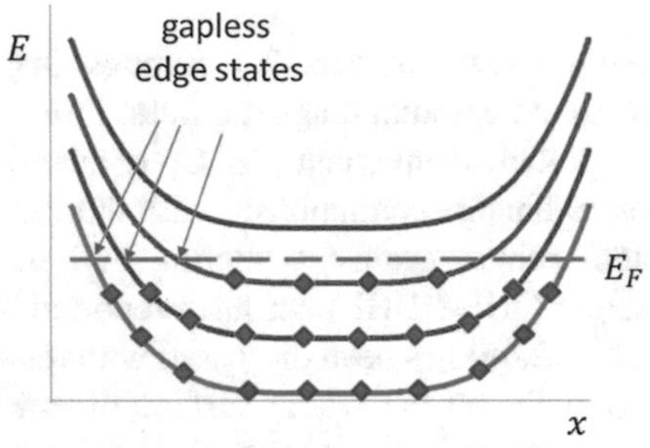

Fig. 2.8 Quantum Hall effect. Landau levels across a 2D sample. Electrons in the interior of the sample (shown as diamonds) do not contribute diagonal current when the Fermi level lies in the energy gap between successive Landau levels

direction. So, C_1 is the winding number that is an integer. The phase $\gamma(k_y)$ is illustrated in Fig. 2.7. The evolution implies the energy gap between filled and empty states that excludes transitions to excited states and preserves the adiabatic flow. As shown in Fig. 2.7, the Chern number discerns dielectrics as having either normal or topologically non-trivial (twisted, Fig. 2.7b) band structure [30].

The IQHE has been explained using various theoretical approaches. One of them explores the relation of the Kubo expression for the Hall conductivity to the TKNN invariant (Chern number) [31] (see [32] for review). Another is the Laughlin-Halperin argument for σ_{xy} quantization due to the spectral flow under insertion of magnetic flux [33, 34].

Simplified illustration of IQHE in Figs. 2.8 and 2.9 shows two-dimensional Landau levels across the sample, where all electrons are localized in the interior and delocalized at the edge.

In an increasing magnetic field, the Landau level is being depopulated until the level becomes completely empty in the interior of the sample. This results in vanishing longitudinal conductivity and quantized Hall resistance in units of h/e. The number of filled localized states in a particular Landau level, shown as diamonds

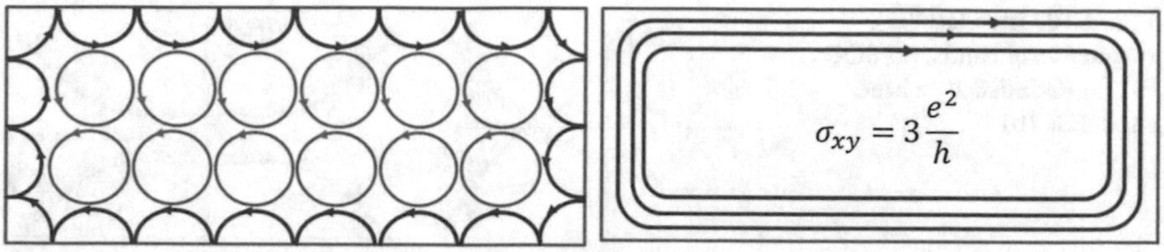

a) b)

Fig. 2.9 (**a**) Semiclassical electron trajectories. Chiral extended edge states (red), localized states in the bulk (blue); (**b**) Non-dissipative edge current. Hall conductivity is quantized in the bulk

in Fig. 2.8, decreases as electrons are pushed out toward the sample's outer boundaries, where gapless extended edge states exist with opposite group velocities at opposite edges (chiral modes). The direction of the chiral mode velocity is determined by the sign of C_1. The absolute value of C_1 is the number of chiral modes, which in IQHE coincides with the number of filled Landau levels. Edge current is shown schematically in Fig. 2.9a, b.

Conditions for the IQHE to occur are more general than the presence of a magnetic flux threading the 2D electron system. A more fundamental property of the Chern insulator is the violation of TR-symmetry. A model Chern insulator and QHE with zero net magnetic field is proposed in [35], where the broken TR-symmetry has been arranged with the complex second-neighbor hopping integral in a honeycomb lattice.

With strong magnetic field and ultralow temperature, the electron–electron interaction becomes dominant and fractional QHE can be observed in which the first Chern number is replaced by fractional numbers. Details on fractional QHE can be found in [36] and the references therein.

2.6.2 *Quantum Anomalous Hall Effect*

As mentioned earlier, the phenomenology of AHE and SHE relies on the common kinetic coefficient (2.8) that indicates the common microscopical mechanism which is at the origin of the effects. In metals, the AHE is not quantized as it relies on the Berry phase picked up on a closed contour on the Fermi surface thus being determined by bulk states.

All arguments regarding 2D Chern insulators and Hall effect quantization can be applied to the QAHE if the spin–orbit interaction is taken into consideration. Quantized Hall effects share the same necessary conditions: the Fermi level should be in the energy gap to suppress the diagonal conductivity in the interior, the TR-symmetry is to be broken to allow chiral states to carry the net edge charge currents, and twisted boundary conditions provide for the non-trivial Berry phase. In QHE, the gaps are in between Landau levels and boundary conditions are twisted by

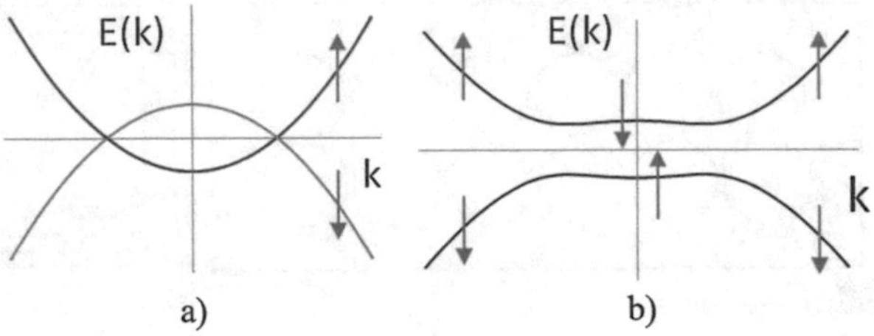

Fig. 2.10 Inverted 2D magnetic subbands, no SOC (**a**); SOC-induced twisted subbands (**b**)

a threading magnetic flux. In the QAHE only the 2D Bloch states, divided by energy gaps, provide for quantization of the Hall conductivity if the Fermi level is in the gap. Then the bulk is a Chern insulator. Instead of the magnetic field being the origin of the Chern insulator, in the QAHE the inverted spectrum with degeneracy lifted by the spin–orbit interaction leads to the quantized Berry phase. Twisted bands in metals are illustrated in Fig. 2.6. In dielectrics, the spin–orbit interaction (SOC)-induced twisted magnetic bands are shown in Fig. 2.10.

So, the integer QAHE is similar to the IQHE in that the quantized part of the Hall conductivity is determined by the edge current which is quantitatively characterized by the Chern number. What discerns QAHE from QHE, aside from the absence of the magnetic field in the QAHE, is the spin–orbit interaction which, along with magnetization, is a source of the anomalous velocity in QAHE, while it is not essential in the QHE.

The objects for QAHE are 2D and thin film dielectrics. Material systems where QAHE has been observed in experiments are *Mn*-doped *CdTe*/*HgTe* quantum well [37] magnetic topological insulators (TI) $Bi_2Se_3(Cr, Fe)$ [38–41], and TI films covered by ferromagnetic materials (see Chap. 3 for details). These material systems demonstrate the necessary conditions for quantized effects to occur: a 2D inverted spectrum and a dielectric phase in the interior of the plane due to the gap between confinement levels in the *CdTe*/*HgTe* quantum well, and the strong exchange field in the TI. One-dimensional chiral edge modes in the QAHE are schematically shown in Fig. 2.11.

Figure 2.11 illustrates the QAHE with a single chiral mode. From the theoretical perspective, multi-chiral edge states ($C_1 > 1$) are not forbidden. So far, however, only one chiral mode has been observed in experiments.

2.6.3 Quantum Spin Hall Effect: $\mathbf{Z_2}$ Topological Index

The Chern number serves well to identify the topologically non-trivial band structure in solids with broken TR-symmetry. If the TR-symmetry is respected, Hall conductivity is absent. That is the case of SHE where the Hall spin current is finite. It is expected that in 2D dielectrics quantum SHE (QSHE) may take place, similarly to QAHE. The QSH state has been proposed and experimentally observed in topological insulators and graphene [42–47]. For example, QSH states were found in the same

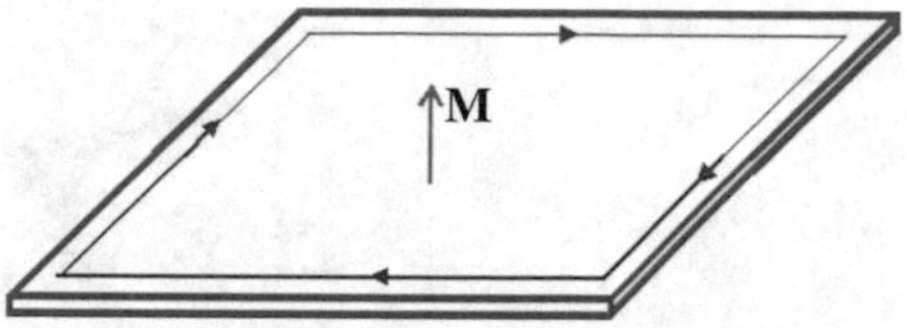

Fig. 2.11 QAHE Chern insulator. Net edge chiral current (red arrows), $C_1 = 1$

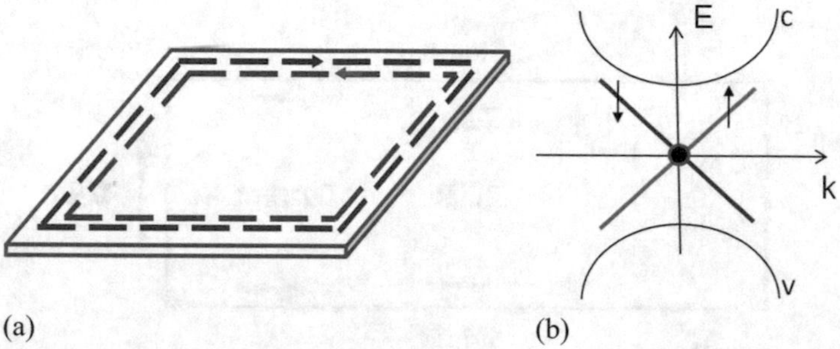

Fig. 2.12 QSHE phase. (**a**) Helical non-dissipative edge charge currents. Red and blue arrows correspond to spin-up and spin-down electrons. In equilibrium, the currents compensate each other. (**b**) Linear-k spectrum of edge gapless spin-momentum locked modes

material systems which demonstrate QAHE: *CdTe/HgTe/CdTe* quantum wells and $Bi_2(Se, Te)_3$. In a quantum well, the 2D plane is in the insulating phase if the Fermi level is placed into the gap between conduction and valence quantization levels.

If Kramers degenerate bands are inverted, the edge states are helical in the sense that they have the locked spin and momentum mutual configuration as discussed in Chap. 1 for a 3D topological insulator (TI). Surface states in TI are gapless Dirac fermions and the gapped spectrum in the interior of the surface stems from the top-bottom mixing term. If this gapped surface spectrum is inverted, the surface is in the QSH state. In quantum wells, the gapped 2D QSH state stems from spatial confinement. In this state, the non-dissipative edge currents of spin-up and spin-down electrons flow along the edge in opposite directions, making the total edge charge current zero, so QHE is absent (see Fig. 2.12).

The energy spectrum for helical states is discussed in Chap. 1 and is shown in Fig. 2.13a. The driving electric field distorts the electron distribution violating the cancellation of currents on opposite edges of the plane illustrated in Fig. 2.12. This creates a spin Hall current in the direction perpendicular to the driving electric field, called QSHE, as shown in Fig. 2.13b.

A gapped spectrum is the necessary condition for QSHE. In 3D TI $Bi_2(Se, Te)_3$, the surface excitations are gapless Dirac fermions. The gap can be open in three ways: by contact with a ferromagnet or a superconductor (proximity effect), and by preparing a TI film thin enough to open the tunneling gap (see Chap. 1). Neither tunneling nor superconductor gaps break the TR-symmetry while the ferromagnetic proximity gap does, making the TI surface an object for QAHE rather than QSHE.

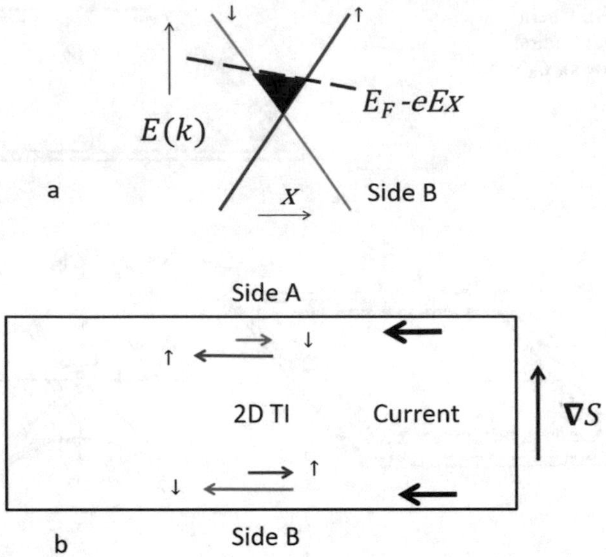

Fig. 2.13 Quantum Spin Hall effect. (**a**) Occupied edge electron states in a biased setup (black area), (**b**) driving current and spin Hall current

Fig. 2.14 Magnetization induced disbalance in helical modes (thick red and thin blue). Black is the resulting chiral current that manifests QAHE

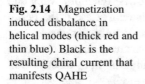

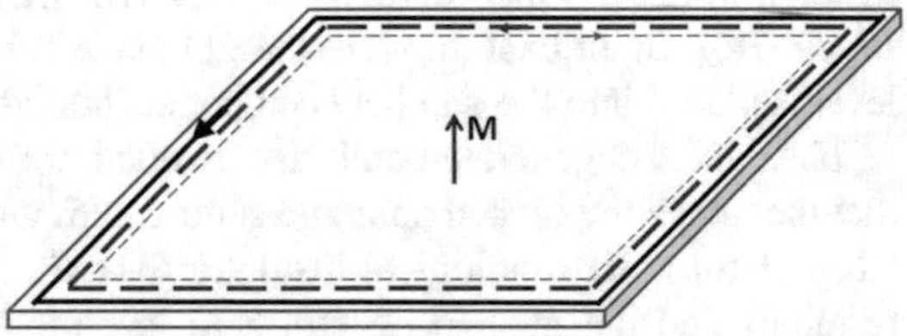

In order to follow the relation between QAHE and QSHE, one may apply magnetization perpendicular to the plane (by the proximity effect) in the setting shown in Fig. 2.12. The cancellation of helical currents becomes broken, quantized Hall conductivity appears, and the picture in Fig. 2.12 changes to carry both helical and chiral states, giving rise to QAHE as illustrated in Fig. 2.14.

The coexistence of the chiral and helical currents in $(Bi, Sb)_{2-x}Cr_xTe_3$ was predicted by first-principal calculations and discussed in relation to experimental data in [48, 49].

In terms of Hall effect quantization, a simplistic picture of the QSHE deals with transverse conductivity in spin-up and spin-down channels characterized by equal Chern numbers of opposite signs, so that the resulting Hall conductivity is zero. This picture implies that spin-up and spin-down eigenstates, one being TR-symmetric to another, can be defined globally, in the whole BZ. However, the eigenvalues of σ_z corresponding to spin-up and spin-down states is not a good quantum number if the

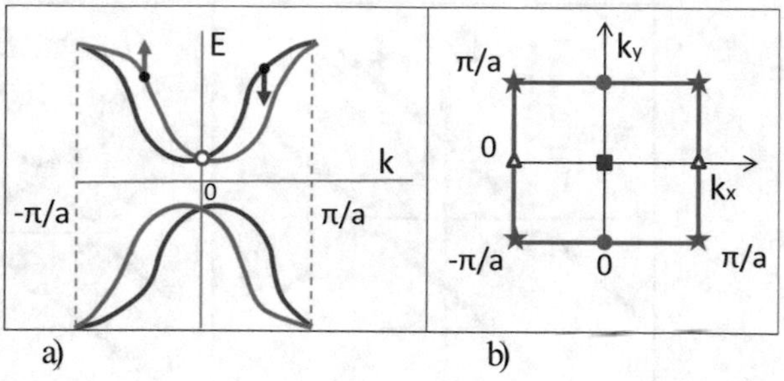

Fig. 2.15 (**a**) Spin-split Kramers doublet branches. (**b**) TRIM in rectangular BZ

spin–orbit coupling takes place. This prevents quantization of the spin Hall current in the QSHE. Even though the Hall effect is absent and the QSHE is not quantized, the QSH phase is not an ordinary dielectric as electron wave functions belong to the Hilbert space of non-trivial topology. The topological invariant that distinguishes the QSH phase from the ordinary insulator is the Z_2 invariant [42, 50]. The invariant is strictly related to Kramers theorem which requires eigenstates to exist as Kramers pairs in the whole BZ. If SI-symmetry holds, Kramers pairs are defined at every point of BZ as the spin branches of the spectrum are two-fold spin degenerate. In a solid without SI-symmetry, the spin–orbit interaction lifts the spin degeneracy at generic k-points and the energy spectrum remains Kramers degenerate: $E_{\uparrow}(\boldsymbol{k}) = E_{\downarrow}(-\boldsymbol{k})$. A generic energy spectrum, extended to the whole BZ, is shown in Fig. 2.15a.

The two example Kramers doublets in Fig. 2.15a are shown at generic points $\pm k$ with black circles, and at $k = 0$. Spin degeneracy is preserved at special points of BZ: there are three k-points where Kramers partners belong to the same momentum: $= -\frac{\pi}{a}, 0, \frac{\pi}{a}$. These are time reversal invariant momenta (TRIM). More generally, TRIM are defined as $\boldsymbol{k} = -\boldsymbol{k} + \boldsymbol{G}$, $\boldsymbol{G}$ being the reciprocal lattice vector. For a 2D rectangular BZ, TRIM are shown in Fig. 2.15b. Four relevant TRIM are shown in the upper right quadrant. TRIM located on opposite sides of the BZ, being separated by $\boldsymbol{G}$, are equivalent and are pictured with the same symbol. Gluing pairs of the same symbols on opposite sides of BZ, one obtains a 2D torus.

Despite the fact that σ_z eigenvalues are not conserved due to the spin–orbit interaction, the degeneracy at TRIM cannot be lifted as it is protected by TR-symmetry (Kramers theorem). The symmetry protection discerns this type of degeneracy from the accidental case discussed above in relation to the AHE.

A single pair of edge modes illustrated in Fig. 2.12a appears as a result of spin-orbit-induced inverted 2D bands. Degeneracy at $\boldsymbol{k} = 0$ is TR-symmetry protected and presents a specific feature of a topologically non-trivial state. If the number of pairs is even, spin-orbit-induced mixing opens the energy gap, making the edges insulating. Along with the insulating interior, this presents an ordinary dielectric (Fig. 2.16a) If the number of pairs is odd, a single gapless Dirac cone remains after the spin–orbit interaction lifts the degeneracy in all accidental $\boldsymbol{k}$-points (Fig. 2.16b). The QSH phase is insensitive to small perturbations that do not violate TR-symmetry.

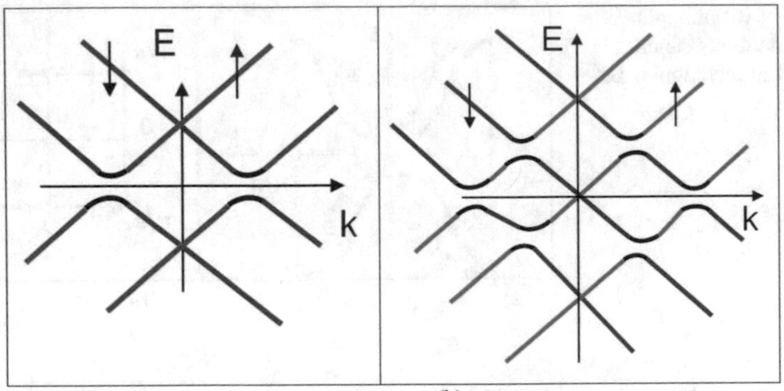

Fig. 2.16 (**a**) Two pairs of edge modes, normal dielectric. (**b**) Three pairs of edge modes. Spin–orbit interaction opens gaps at all accidental $\boldsymbol{k}$-points. Degeneracy at $\boldsymbol{k} = 0$ and other TRIM (not shown) is protected

As the Chern number counts the number of edge states in Chern insulators, the Z_2 invariant counts the number of pairs of counterpropagating states in TR-symmetric dielectrics. The number of pairs is equal to (0 or 1) mod 2, so the topological index $Z_2 = 0, 1$ correspond to trivial and QSH phases, respectively.

TR-symmetry is a dominant factor that discerns the QSH state from the Chern insulator, so the Z_2 index should be related to the properties of a time reversal operator acting in the subspace of occupied bands. The time reversal operator acting on a single particle with spin $s = 1/2$ can be represented as

$$\Theta = -i\sigma_y K = \begin{pmatrix} 0 & -1 \\ 1 & 0 \end{pmatrix} K, \tag{2.49}$$

where σ_y is the Pauli matrix and K is the operator of complex conjugation. Acting on pure spin states $\begin{pmatrix} 1 \\ 0 \end{pmatrix}$ and $\begin{pmatrix} 0 \\ 1 \end{pmatrix}$, the operator Θ flips the spin direction:

$$\Theta \begin{pmatrix} 1 \\ 0 \end{pmatrix} = \begin{pmatrix} 0 \\ 1 \end{pmatrix}, \quad \Theta \begin{pmatrix} 0 \\ 1 \end{pmatrix} = -\begin{pmatrix} 1 \\ 0 \end{pmatrix}, \quad \Theta^2 = 1. \tag{2.50}$$

The basis we chose in the effective surface four-band model (Chap. 1, 1.38) dictates the form of the time reversal operator as

$$\Theta = \begin{pmatrix} 0 & 0 & -1 & 0 \\ 0 & 0 & 0 & -1 \\ 1 & 0 & 0 & 0 \\ 0 & 1 & 0 & 0 \end{pmatrix} K. \tag{2.51}$$

The action of the time reversal operator on Bloch amplitudes (1.41) can be expressed as

$$\Theta C_{\uparrow}(k) = \sqrt{\frac{Bk^2 - \Delta_S/2 + R_{\uparrow}}{4\,R_{\uparrow}}} \begin{pmatrix} \dfrac{(-\tilde{A}_2 k + \tilde{V}_{AS})\, i \exp(-i\varphi)}{Bk^2 - \Delta_S/2 + R_{\uparrow}} \\ i\exp(-i\varphi) \\ \dfrac{\tilde{V}_{AS} - \tilde{A}_2 k}{Bk^2 - \Delta_S/2 + R_{\uparrow}} \\ 1 \end{pmatrix} = C_{\downarrow}(-k),$$

$$\Theta C_{\downarrow}(k) = -\sqrt{\frac{Bk^2 - \Delta_S/2 + R_{\downarrow}}{4\,R_{\downarrow}}} \begin{pmatrix} -\dfrac{(\tilde{A}_2 k + \tilde{V}_{AS})\, i \exp(-i\varphi)}{Bk^2 - \Delta_S/2 + R_{\downarrow}} \\ -i\exp(-i\varphi) \\ \dfrac{\tilde{A}_2 k + \tilde{V}_{AS}}{Bk^2 - \Delta_S/2 + R_{\downarrow}} \\ 1 \end{pmatrix} = -C_{\uparrow}(k),$$

$$\Theta V_{\uparrow}(k) = \sqrt{\frac{\Delta_S/2 - Bk^2 + R_{\uparrow}}{4R_{\uparrow}}} \begin{pmatrix} -\dfrac{(\tilde{A}_2 k - \tilde{V}_{AS})\, i \exp(-i\varphi)}{\Delta_S/2 - Bk^2 + R_{\uparrow}} \\ i\exp(-i\varphi) \\ \dfrac{\tilde{A}_2 k - \tilde{V}_{AS}}{\Delta_S/2 - Bk^2 + R_{\uparrow}} \\ 1 \end{pmatrix} = V_{\downarrow}(-k),$$

$$\Theta V_{\downarrow}(k) = -\sqrt{\frac{\Delta_S/2 - Bk^2 + R_{\downarrow}}{4R_{\downarrow}}} \begin{pmatrix} \dfrac{(\tilde{A}_2 k + \tilde{V}_{AS})\, i \exp(-i\varphi)}{\Delta_S/2 - Bk^2 + R_{\downarrow}} \\ -i\exp(-i\varphi) \\ \dfrac{\tilde{A}_2 k + \tilde{V}_{AS}}{\Delta_S/2 - Bk^2 + R_{\downarrow}} \\ 1 \end{pmatrix} = -V_{\uparrow}(-k). \quad (2.52)$$

Using (2.52), it is straightforward to obtain the matrix overlap of Kramers conjugate partners

$$w(\boldsymbol{k}) = \langle i|\Theta|j\rangle = \begin{pmatrix} 0 & 0 & a_{13} & a_{14} \\ 0 & 0 & a_{23} & a_{24} \\ -a_{13} & -a_{23} & 0 & 0 \\ -a_{14} & -a_{24} & 0 & 0 \end{pmatrix}, \quad i,j = (1,2,3,4) \to (V\uparrow, C\uparrow, V\downarrow, C\downarrow),$$

$$a_{13} = \frac{\tilde{A}_2^2 k^2 - \tilde{V}_{AS}^2 - \left(\Delta_S/2 - Bk^2 + R_{\uparrow}\right)\left(\Delta_S/2 - Bk^2 + R_{\downarrow}\right)}{2\sqrt{R_{\uparrow}\, R_{\downarrow}}\sqrt{\Delta_S/2 - Bk^2 + R_{\uparrow}}\sqrt{\Delta_S/2 - Bk^2 + R_{\downarrow}}},$$

$$a_{14}=\frac{\tilde{V}_{AS}^2-\tilde{A}_2^2k^2-\left(\Delta_S/2-Bk^2+R_\uparrow\right)\left(Bk^2-\Delta_S/2+R_\downarrow\right)}{2\sqrt{R_\uparrow R_\downarrow}\sqrt{\Delta_S/2-Bk^2+R_\uparrow}\sqrt{Bk^2-\Delta_S/2+R_\downarrow}},$$

$$a_{23}=\frac{\tilde{V}_{AS}^2-\tilde{A}_2^2k^2-\left(Bk^2-\Delta_S/2+R_\uparrow\right)\left(\Delta_S/2-Bk^2+R_\downarrow\right)}{2\sqrt{R_\uparrow R_\downarrow}\sqrt{Bk^2-\Delta_S/2+R_\uparrow}\sqrt{\Delta_S/2-Bk^2+R_\downarrow}},$$

$$a_{24}=\frac{\tilde{A}_2^2k^2-\tilde{V}_{AS}^2-\left(Bk^2-\Delta_S/2+R_\uparrow\right)\left(Bk^2-\Delta_S/2+R_\downarrow\right)}{2\sqrt{R_\uparrow R_\downarrow}\sqrt{Bk^2-\Delta_S/2+R_\uparrow}\sqrt{Bk^2-\Delta_S/2+R_\downarrow}}. \tag{2.53}$$

We assume that the chemical potential is located in the tunneling gap and express the overlap of occupied (valence) bands as a submatrix:

$$w(\boldsymbol{k})=\begin{pmatrix}0 & a_{13}\\ -a_{13} & 0\end{pmatrix}, \tag{2.54}$$

where $\det[w(\boldsymbol{k})]=P_{\boldsymbol{k}}^2=a_{13}^2,\ P_{\boldsymbol{k}}$ is the Pfaffian to the matrix (2.54). The Pfaffian plays a special role in the identification of the non-trivial insulator state as the Berry connection is determined by the phase of the Pfaffian, $\boldsymbol{A}(\boldsymbol{k})=i\boldsymbol{\nabla}_{\boldsymbol{k}}\, log\ P_{\boldsymbol{k}}$ [50] and one expects singularity in the Berry connection if the energy spectrum allows an accidental degeneracy point (see (2.26)). Singularity in the Berry connection means a zero Pfaffian. Below we follow zeroes of $P_{\boldsymbol{k}}$. In terms of obstruction (see the discussion that follows (2.32)), zero $P_{\boldsymbol{k}}$ implies that one cannot define the phase of the Bloch amplitude in the whole BZ because of the constraint introduced by Kramers theorem [51]. At TRIM, spin-up and spin-down states are physically equivalent, so $|\ P_{\boldsymbol{K}_i}\ |=1$, $\boldsymbol{K}_i$ runs over four TRIM in square BZ. If for some choice of parameters in the Hamiltonian the Pfaffian $P_{\boldsymbol{k}}$ becomes zero at an accidental point in BZ, $\boldsymbol{k}^*$, another zero is located in $-\boldsymbol{k}^*$ as zeroes come by pairs at Kramers conjugate momenta. So, the number of zeroes in the whole BZ is always equal to $0 \mod 2$. Thus, we need only half the BZ to count the number of pairs of zeroes I. One may calculate I as the Berry phase or winding number that the phase of $P_{\boldsymbol{k}}$ acquires while $\boldsymbol{k}$ runs over the loop enclosing the half-BZ:

$$I=\frac{1}{2\pi i}\oint_C d\boldsymbol{k}\ \boldsymbol{\nabla}_{\boldsymbol{k}}\log(P_{\boldsymbol{k}}+i\delta), \tag{2.55}$$

where the integration path in the square lattice is shown in Fig. 2.17.

An integration loop connects TRIM, where $|P_{\boldsymbol{K}_i}|=1$. If one of four $P_{\boldsymbol{K}_i}$ has a sign different from the other three, it means that $P_{\boldsymbol{k}}$ changes sign at least twice in two zeroes along the contour in Fig. 2.17. That, in turn, gives $I=1$. This number is called

Fig. 2.17 Path to follow the winding of the phase of $P_{\boldsymbol{k}}$. Four TRIM points: Γ, B, C, D. Example: zeroes of $P_{\boldsymbol{k}}$ along the path on k_x axis

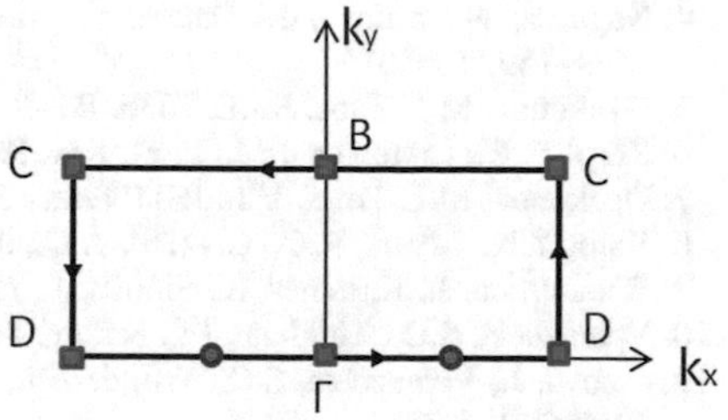

the Z_2 invariant. Even and odd winding numbers ($I = 0$, $1\ mod\ 2$) correspond to the normal dielectric and the QSH phase, respectively. In general, the feature of a topologically non-trivial state is recognized by a negative sign of the product $\prod_{\boldsymbol{K}_i} \boldsymbol{P}_{\boldsymbol{K}_i}$, where $\boldsymbol{P}_{\boldsymbol{K}_i}$ is calculated from the overlap matrix of occupied bands.

In our example (2.54) and (2.55), we are unable to control all TRIM as the model is designed to describe the states and eigenvalues in the vicinity of Γ-point in BZ. In the continuous limit $G = 2\pi/a \to \infty$, all three TRIM B, C, D move to infinity and the change of the sign of $P_{\boldsymbol{k}}$ can be followed on a k-half-axis $(0, \infty)$. Opposite signs of $P_{\boldsymbol{0}}$ and P_{∞} manifest the QSH state. If the vertical bias across the film is zero, $\tilde{V}_{AS} = 0$,

$$P_{\boldsymbol{k}} = \frac{Bk^2 - \Delta_S/2}{\sqrt{\left(Bk^2 - \Delta_S/2\right)^2 + \tilde{A}_2^2 k^2}}. \tag{2.56}$$

It is easy to check that $P_0 = -\ \text{sign}\ (\Delta_S)$, $P_{\infty} = \text{sign}\ (B)$, so $B\Delta_S > 0$ is the condition for the band inversion and the QHS state to occur. In this simple case, the topological index I can be expressed as $(-1)^I = \text{sign}\ [-B\Delta_S]$.

If both TR and SI symmetries hold, the overlap matrix $w(\boldsymbol{k})$ is composed of matrix elements $\langle i| P\Theta |j\rangle$, where P is the inversion operator. In this case, all TRIM states have certain parity and the feature of the QSH state is the negative sign of the product $\prod_N \prod_{\boldsymbol{K}_i} \xi(\boldsymbol{K}_i)$, where $\xi(\boldsymbol{K}_i)$ is the parity eigenvalue of an occupied band in point $\boldsymbol{K}_i$, and N is the number of occupied bands.

References

1. Hasan, M.Z., Kane, C.L.: Rev. Mod. Phys. **82**, 3045–3067 (2010)
2. Anselm, A.: Introduction to Semiconductor Theory. Prentice Hall, Englewood Cliffs, NJ (1981)
3. Wolf, C.M., Holonyak Jr., N., Stillman, G.E.: Physical Properties of Semiconductors. Prentice Hall, Englewood Cliffs, NJ (1989)

4. Nagaosa, N., Sinova, J., Onoda, S., MacDonald, A.H., Ong, N.P.: Rev. Mod. Phys. **82**, 1539–1592 (2010)
5. Dyakonov, M.I.: Proc. SPIE. **7036**, R1–R14 (2008)
6. Karplus, R., Luttinger, J.M.: Phys. Rev. **95**, 1154 (1954)
7. Dyakonov, M.I., Perel, V.I.: JETP Lett. **13**, 467–469 (1971)
8. Kato, Y.K., Myers, R.C., Gossard, A.C., Awschalom, D.D.: Science. **306**, 1910 (2004)
9. Wunderlich, J., Kaestner, B., Sinova, J., Jungwirth, T.: Phys. Rev. Lett. **94**, 047204 (2005)
10. Valenzuela, S.O., Tinkham, M.: Nature. **442**, 176 (2006)
11. Sinova, J., Valenzuela, S.O., Wunderlich, J., Back, C.H., Jungwirth, T.: Rev. Mod. Phys. **87**, 1213 (2015)
12. Berry, M.V.: Proc. R. Soc. Lond. **A392**, 45 (1984)
13. Sakurai, J.J.: Modern Quantum Mechanics. Addison Wesley, Reading, MA (1993)
14. Chern, S.S.: Ann. Math. **46**, 674–684 (1945)
15. Nakahara, M.: Geometry, Topology and Physics. Taylor and Francis, New York & London (2003)
16. Hatsugai, Y.: J. Physical Soc. Japan. **75**, 123601 (2006)
17. Zak, J.: Phys. Rev. Lett. **62**, 27472750 (1989)
18. Resta, R.: J. Phys. Condens. Matter. **12**, R107–R143 (2000)
19. Sundaram, G., Niu, Q.: Phys. Rev. **B59**, 14915 (1999)
20. Xiao, D., Chang, M.C., Niu, Q.: Rev. Mod. Phys. **82**, 1959–2007 (2010)
21. "Berry's Phase and Geometric Quantum Distance: Macroscopic Polarization and Electron Localization". https://cds.cern.ch/record/893386, EPFL, Lausanne (2000)
22. Adams, E.N., Blount, E.I.: J. Phys. Chem. Solid. **10**, 286 (1959)
23. Blount, E.I.: In: Seitz, F., Turnbull, D. (eds.) Solid State Physics, vol. 13, p. 305. Academic Press, New York (1962)
24. Marder, M.: Condensed Matter Physics. Wiley, New York (2000)
25. Shindou, R., Imura, K.: Nucl. Phys. **B720**, 399–435 (2005)
26. Onoda, M., Nagaosa, N.: J. Phys. Soc. Japan. **71**(1), 19–22 (2002)
27. Sinova, J., Culcer, D., Nui, Q., Sinitsyn, N.A., Jungwirth, T., MacDonald, A.H.: Phys. Rev. Lett. **92**(12), 126603 (2004)
28. Yao, Y., Fang, Z.: Phys. Rev. Lett. **95**, 156601 (2005)
29. von Klitzing, K., Dorda, G., Pepper, M.: Phys. Rev. Lett. **45**, 494 (1980)
30. Weng, H., Yu, R., Hu, X., Dai, X., Fang, Z.: arXiv:1508.02967v3 [cond-mat.mes-hall] (2015)
31. Thouless, J.D., Kohmoto, M., Nightingale, P., den Nijs, M.: Phys. Rev. Lett. **49**, 405 (1982)
32. Ando, Y.: J. Phys. Soc. Jpn. **82**, 102001 (2013)
33. Laughlin, R.B.: Phys. Rev. **B23**, 5632 (1981)
34. Halperin, B.I.: Phys. Rev. **B25**, 2185 (1982)
35. Haldane, F.D.M.: Phys. Rev. Lett. **61**, 2015 (1988)
36. Jain, J.K.: Adv. Phys. **41**, 105 (1992)
37. Liu, C.X., Qi, X.L., Dai, X., Fang, Z., Zhang, S.C.: Phys. Rev. Lett. **101**, 146802 (2008)
38. Yu, R., Zhang, W., Zhang, H.J., Zhang, S.C., Dai, X., Fang, Z.: Science. **329**, 61 (2010)
39. Chang, C.Z., Zhang, J., Feng, X., Shen, J., Zhang, Z., Guo, M., Li, K., Ou, Y., Wei, P., Wang, L.L., Ji, Z.Q., Feng, Y., Ji, S., Chen, X., Jia, J., Dai, X., Fang, Z., Zhang, S.C., He, K., Wang, Y., Lu, L., Ma, X.C., Xue, Q.K.: Science. **340**, 167–170 (2013)
40. Checkelsky, J.G., Yoshimi, R., Tsukazaki, A., Tahakashi, K.S., Kozuka, Y., Falson, J., Kawasaki, M., Tokura, Y.: Nat. Phys. **10**, 731–736 (2014)
41. Kou, X., Guo, S.T., Fan, Y., Pan, L., Lang, M., Jiang, Y., Shao, Q., Nie, T., Murata, K., Tang, J., Wang, Y., He, L., Lee, T.K., Lee, W.L., Wang, K.L.: Phys. Rev. Lett. **113**, 137201 (2014)
42. Kane, C.L., Mele, E.J.: Phys. Rev. Lett. **95**, 146802 (2005)
43. Bernevig, B.A., Zhang, S.C.: Phys. Rev. Lett. **96**, 106802 (2006)
44. Bernevig, B.A., Hughes, T.L., Zhang, S.C.: Science. **314**, 1757 (2006)
45. Konig, M., Wiedmann, S., Brüne, C., Roth, A, Buhmann, H., Molenkamp, L.W., Qi, X.-L., Zhang, S.-C.: Science. **318**, 766 (2006)

46. Roth, A., Brüne, C., Buhmann, H., Molenkamp, L.W., Maciejko, J., Qi, X.-L., Zhang, S.-C.: Science. **325**, 294 (2009)
47. Kane, C.L., Mele, E.J.: Phys. Rev. Lett. **95**, 226801 (2005)
48. Wang, J., Lian, B., Zhang, H., Zhang, S.C.: Phys. Rev. Lett. **111**, 086803 (2013)
49. Wang, J., Lian, B., Zhang, S.C.: arXiv:1409.6715v4 [cond-mat.mes-hall] (2015)
50. Fu, L., Kane, C.L.: Phys. Rev. **B76**, 045302 (2007)
51. Fruchart, M., Carpentier, D.: arXiv:1310.0255v2 [cond-mat.mes-hall] (2013)

Chapter 3
Magnetic Field and Ferromagnetic Proximity Effects

The device applications of topological insulators, let's say transistors, require a gapped electron spectrum. The gap also provides the conditions for the quantum anomalous Hall effect [1] and topological magnetoelectric effects [2, 3]. The surface energy spectrum of a TI film, discussed in Chap. 1, is gapped as long as the states on opposite surfaces overlap. The overlap opens the tunneling gap in the otherwise massless electron spectrum when the Bi_2Se_3 film thickness becomes less than six quintuple layers [4]. If the thickness increases, the gap tends to zero and each surface independently carries 2D gapless Dirac fermions. In a thick film, the tunneling gap is negligibly small, and the energy gap on the surface can be created by breaking the time reversal symmetry with an out-of-plane magnetic field. This setting induces Landau levels and creates conditions for the QHE. Without an external magnetic field the loss of the time reversal symmetry in TI might be caused by magnetic doping with transition metal atoms (Fe, Mn, Cr, V) to the extent that the surface becomes ferromagnetic [5–9]. Also, the proximity effect from the ferromagnetic insulator (FM) creates out-of-plane magnetization of surface electrons in TI, thus creating conditions for the QAHE [10–12]. The experimental observation of the QAHE in an FM/TI bilayer is considered proof of the proximity ferromagnetism. For example, the Bi_2Se_3/EuS heterostructure reveals proximity-induced high temperature magnetization in the Bi_2Se_3 in a 20 $\dot{A}$ -deep near-interface region. What is essential for the development of TI-based spintronic devices is that the surface helical electrons affect the magnetization dynamics in an adjacent ferromagnet by the spin-transfer torque, that is, the transfer of spin angular momentum between surface electrons and the magnetization in the ferromagnetic layer [13–15].

In this chapter, we consider the exchange field-induced gapped spectrum in 2D- and 3D- TI and discuss the role of in-plane and out-of-plane magnetic fields on the topological properties of the surface.

V. Litvinov, *Magnetism in Topological Insulators*,
https://doi.org/10.1007/978-3-030-12053-5_3

3.1 Magnetic Energy Gap

We start with the Hamiltonian of the surface states (Chap. 1, (1.38)) which includes Zeeman energy, $m_z = JM_z$, originating from out-of-plane proximity magnetization in an adjacent ferromagnetic (FM) layer, J is the exchange interaction constant:

$$\tilde{H} = \begin{pmatrix} \frac{\Delta_S}{2} - Bk^2 + m_z & \tilde{V}_{AS} & 0 & i\tilde{A}_2 k_- \\ \tilde{V}_{AS} & -\frac{\Delta_S}{2} + Bk^2 + m_z & i\tilde{A}_2 k_- & 0 \\ 0 & -i\tilde{A}_2 k_+ & \frac{\Delta_S}{2} - Bk^2 - m_z & \tilde{V}_{AS} \\ -i\tilde{A}_2 k_+ & 0 & \tilde{V}_{AS} & -\frac{\Delta_S}{2} + Bk^2 - m_z \end{pmatrix}, \tag{3.1}$$

Both parameters Δ_S and B originate from the quantum tunneling between electron states residing on opposite surfaces and tend to zero if the film thickness exceeds six quintuple layers. All three components of magnetization are needed to describe the magnetic anisotropy in a ferromagnetic layer and will be discussed in Sect. 3.7 of this chapter.

3.1.1 Independent Surfaces

Below we analyze the eigenvalues of the Hamiltonian (3.1) that account for out-of-plane magnetization only, and we consider various settings representing surface states in 2D and 3D topological regimes. In order to get the spectrum of a single surface from the Hamiltonian (3.1), we neglect the tunneling ($\Delta_S = B = 0$) and external voltage ($\tilde{V}_{AS} = 0$). This means that we consider a thick film with two independent and electrically identical surfaces. If no proximity ferromagnetism is taken into account ($\boldsymbol{m} = 0$), one obtains a massless spin-resolved spectrum that is double degenerate with the surface index (top and bottom). Each of the two surfaces carries electrons with spectrum $E = \pm\tilde{A}_2 k$ shown in Fig. 3.1a. The proximity magnetization along the c-axis modifies the spectrum as $E = \pm\sqrt{\tilde{A}^2 k^2 + m_z^2}$, thus creating the energy gap of $E_g = 2 \mid m_z|$, as illustrated in Fig. 3.1b.

The existence of the linear spectrum relies on spin–orbit interaction $(\tilde{A}_2 \neq 0)$. The magnetization- induced energy gap is essential for realization of the QAHE. Observation of the QAHE may serve as experimental evidence of a good quality interface between the TI and a ferromagnetic layer. The magnitude of the magnetic gap can be as large as 100 meV [16].

In a real device, the film is grown on a substrate which creates the potential difference between the top and the bottom surfaces, $\tilde{V}_{AS}$. This shifts branches of

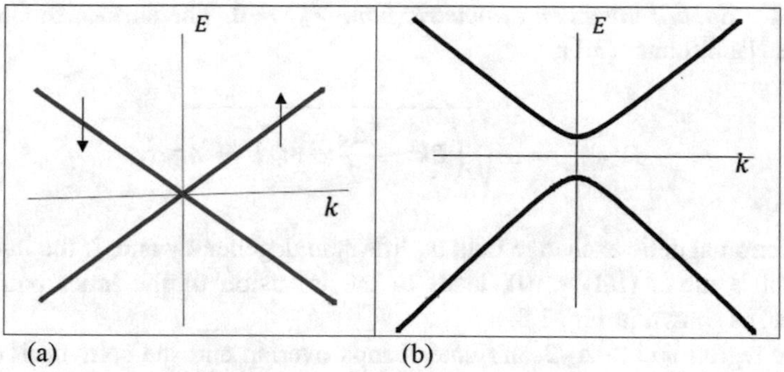

Fig. 3.1 (**a**) Single surface spectrum, $m = 0$. (**b**) Magnetic gap, $m_z \neq 0$

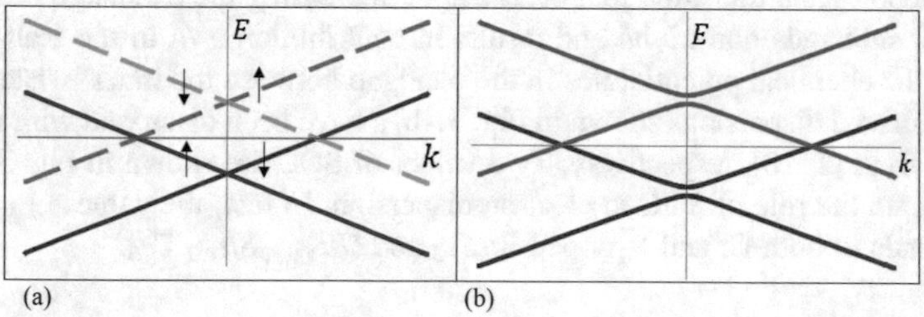

Fig. 3.2 Independent surfaces. Rashba spin-split spectrum, (**a**) $m_z = 0$. (**b**) Magnetic gaps, $m_z \neq 0$

electrons residing on opposite surfaces, so the surface electrons acquire dispersion, expressed as

$$E = \begin{cases} \pm\tilde{A}_2 k \pm \tilde{V}_{AS} & m_z = 0, \\ \pm\tilde{V}_{AS} \pm \sqrt{m_z^2 + \tilde{A}^2 k^2} & m_z \neq 0, \end{cases} \tag{3.2}$$

and illustrated in Fig. 3.2.

In the absence of a magnetic field, the gapped spectrum (3.2) provides the condition for QAHE on the FM/TI interface due to the proximity to the FM overlayer. In a QHE regime, an external magnetic field takes part in the setting instead of FM, and the gapless spectrum, shown in Fig. 3.2a, favors the configuration in which two opposite surfaces may cancel the QHE [17].

3.2 Proximity-Induced Topological Phase Transition

In a thin film, the overlap of the top-bottom wave functions results in a tunneling gap and parabolic in-plane dispersion, $\Delta_S \neq 0, B \neq 0$. Several regimes define the electron dispersion and conditions upon which bands become topologically non-trivial, giving rise to QAHE.

1. *Spatial inversion symmetric film,* $\tilde{V}_{AS} = 0$. The surface modes follow from the Hamiltonian (3.1):

$$E_{c,v,\uparrow,\downarrow} = \pm\sqrt{\left(Bk^2 - \frac{\Delta_S}{2} \pm m_z\right)^2 + \tilde{A}_2^2 k^2}. \tag{3.3}$$

The ferromagnetic exchange field m_z lifts spin degeneracy and, if the initial band alignment is direct ($B\Delta_S < 0$), leads to the inversion of the inner pair of spin subbands, as shown in Fig. 3.3.

In the region $|m_z| > \Delta_S/2$, the inner bands overlap and the spin–orbit coupling (SOC) opens the gap at $k = \pm\sqrt{\Delta_S/2B}$ as illustrated in Fig. 3.4. Parameters: $\Delta_S = 0.1$ eV, $|B| = 30$ eVÅ^2.

The topological transition that occurs upon increasing the exchange field makes the inner subbands non-trivial and results in contribution e^2/h to the Hall conductivity if the chemical potential lies in the bandgap between the inner twisted bands. The non-trivial dispersions shown in Fig. 3.4b, c have been discussed with regard to the QAHE in [1, 18], respectively. Two values of SOC are shown in Fig. 3.4b, c to demonstrate the role of SOC in electron dispersion. In fact, the value $\tilde{A}_2 = 3$ eVÅ corresponds to both Cr and V-doped Bi_2Se_3 and $(Bi_{1-x}Sb_x)_2Te_3$.

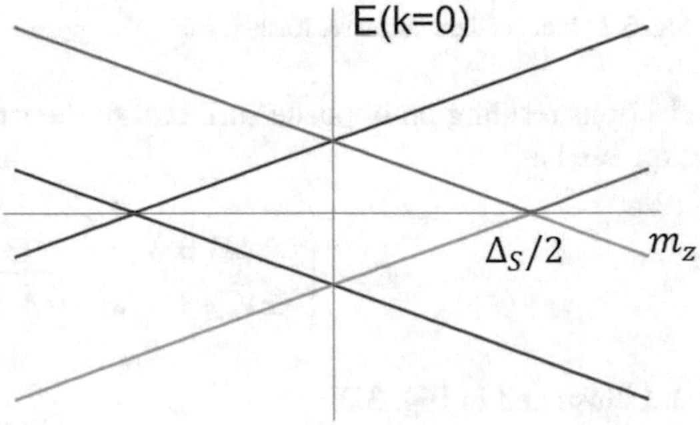

Fig. 3.3 Energy levels (3.3) at $k = 0$. Inner subbands are inverted at $|m_z| > \Delta_S/2$

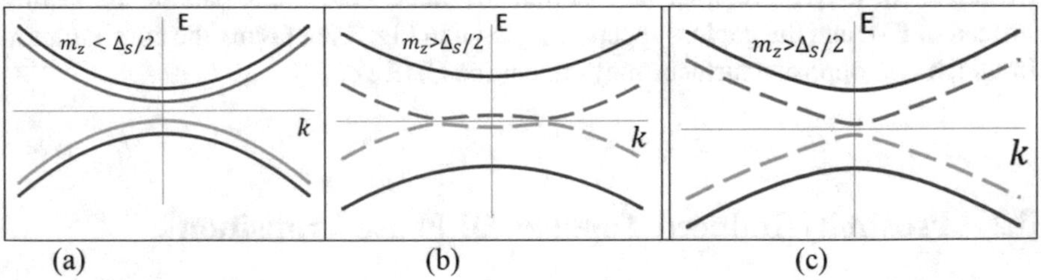

Fig. 3.4 Thin film energy spectrum. (**a**) Direct band alignment $\left(B\Delta_S < 0,\ \tilde{A}_2 = 0\right)$, (**b**) $\tilde{A}_2 = 0.3$ eVÅ .The inner pair of bands (dashed) becomes inverted when the exchange field m_z increases; (**c**) same as (**b**), $\tilde{A}_2 = 3$ eVÅ

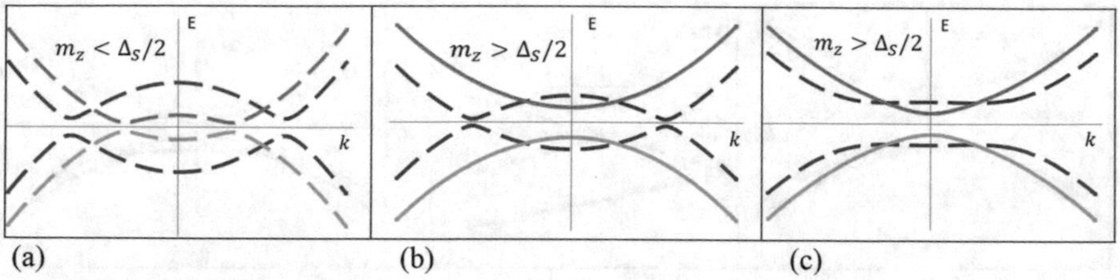

Fig. 3.5 (**a**) Inverted band alignment $(B\Delta_S > 0)$, $\tilde{A}_2 = 0.3$ eVÅ, (**b**) released inversion in the inner pair of bands (solid lines) upon increasing exchange field, $\tilde{A}_2 = 0.3$ eVÅ. (**c**) same as (**b**), $\tilde{A}_2 = 3$ eVÅ

If the initial subbands are inverted $(B\Delta_S > 0)$, the increasing exchange field releases the inversion of inner bands at $|m_z| > \Delta_S/2$, so the inner band alignment becomes direct, as illustrated in Fig. 3.5b, c.

The band arrangement, rendered in Fig. 3.5b, is recognized as the QAHE phase [1], because the inverted subbands (dashed lines) carry nonzero Chern numbers. Basically, no matter whether the initial bands are inverted, as in Fig. 3.5a, or not, as in Fig. 3.4a, the exchange field may invert the inner pair of bands, thus providing the conditions for the QAHE (Fig. 3.4b, c). What is mandatory for the QAHE to occur is that the minimum energy gap is formed by the inverted bands. In a real material system $Bi_2Se_3(Cr, Fe)$, the parameters correspond to the case illustrated in Fig. 3.5c where the minimum gap is formed by the subbands in the normal positions (not inverted, solid lines). These subbands carry zero Chern numbers, so no quantization of the AHE may take place. However, quantization may appear if inversion asymmetry is taken into account. The asymmetry affects the topology of the actual bands and results in the QAHE, as discussed below.

2. *Spatial inversion asymmetric film*, $\tilde{V}_{AS} \neq 0$. The nonzero asymmetric potential is proportional to the top-bottom potential difference originated either from the built-in potential or applied vertical bias, or both. The dispersion of the surface modes following from (3.1) is given as

$$E_{c,v\uparrow}(k) = \begin{cases} \pm\sqrt{\left(\frac{\Delta_S}{2} - Bk^2\right)^2 + \left(\tilde{A}_2 k - \tilde{V}_{AS}\right)^2}, & m_z = 0 \\ \pm\sqrt{\left(\frac{\Delta_S}{2} - Bk^2\right)^2 + m_z^2 + \tilde{A}_2^2 k^2 + \tilde{V}_{AS}^2 - r(k)}, & m_z \neq 0 \end{cases},$$

$$E_{c,v\downarrow}(k) = \begin{cases} \pm\sqrt{\left(\frac{\Delta_S}{2} - Bk^2\right)^2 + \left(\tilde{A}_2 k + \tilde{V}_{AS}\right)^2}, & m_z = 0 \\ \pm\sqrt{\left(\frac{\Delta_S}{2} - Bk^2\right)^2 + m_z^2 + \tilde{A}_2^2 k^2 + \tilde{V}_{AS}^2 + r(k)}, & m_z \neq 0 \end{cases},$$

$$r(k) = 2\sqrt{m_z^2\left(\frac{\Delta_S}{2} - Bk^2\right)^2 + \left(m_z^2 + \tilde{A}_2^2 k^2\right)\tilde{V}_{AS}^2}. \tag{3.4}$$

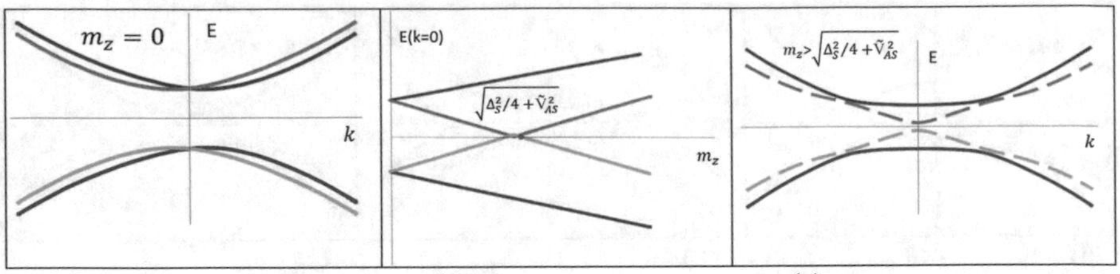

(a) (b) (c)

Fig. 3.6 (**a**) Rashba spin-splitting, $\tilde{A}_2 = 3$ eVÅ, $\tilde{V}_{AS} = 0.01$ eV. (**b**) Swap of energy levels upon increasing the exchange field. (**c**) The accidental degeneracy shown in Fig. 3.5c at finite k is lifted by the asymmetry potential

The vertical bias ($\tilde{V}_{AS} \neq 0$) and SOC $\left(\tilde{A}_2 \neq 0\right)$ lift the spin degeneracy (Rashba spin-splitting) as illustrated in Fig. 3.6a. The level positions $E_{c\uparrow}$ and $E_{v\downarrow}$ at $k = 0$ swap at $m_z = \sqrt{\Delta_S^2/4 + \tilde{V}_{AS}^2}$ as shown in Fig. 3.6b. As we consider the setting with initially inverted bands, $B\Delta_S > 0$, the swap means the release of the band inversion between the inner pair of subbands.

The asymmetry potential causes band anti-crossing (compare Figs. 3.5c and 3.6c) so that the topological index of inner bands becomes non-trivial [18]. The process is similar to the AHE in metals, where a non-trivial topology appears due to bands anti-crossing as illustrated in Chap. 2, Figs. 2.5 and 2.6. So, the initially topological phase ($B\Delta_S > 0$) of $Bi_2Se_3(Cr, Fe)$ is destroyed by a strong proximity field and then the QAHE phase reinstates manifesting the topological phase transition induced by a vertical bias.

A simple example that illustrates how anti-crossing of the levels generates the non-trivial topology and the QAHE in the gapped TI is the subject of the next section.

3.3 Quantum Anomalous Hall Effect

As an example, we consider the chemical potential located in the energy gap between a pair of inner bands inverted by a strong exchange field, as shown in Fig. 3.4c. The Hamiltonian of this subsystem is a submatrix of the full expression (3.1):

$$H_{\mathbf{k}} = \begin{pmatrix} \frac{\Delta_S}{2} - Bk^2 - m_z & i\tilde{A}_2 k_- \\ -i\tilde{A}_2 k_+ & -\frac{\Delta_S}{2} + Bk^2 + m_z \end{pmatrix}. \tag{3.5}$$

We assume direct initial band alignment in the absence of the exchange field, $B\Delta_S < 0$ (see Fig. 3.4). Eigenvalues of the Hamiltonian (3.5) are given as

$$E_{c,v} = \pm E_k, \quad E_k = \sqrt{\left(Bk^2 - \frac{\varepsilon_g}{2}\right)^2 + \tilde{A}_2^2 k^2}, \quad \varepsilon_g = 2m_z - \Delta_S. \tag{3.6}$$

The notations in (3.6) imply that the spectrum is inverted under the condition $B\varepsilon_g > 0$. The Bloch amplitudes in bands $\pm E_k$ normalized to unity are given as

$$u_- = \frac{\tilde{A}_2\, k}{\sqrt{\tilde{A}_2^2 k^2 + \left(Bk^2 - \varepsilon_g/2 + E_k\right)^2}} \begin{pmatrix} i\dfrac{\varepsilon_g/2 - Bk^2 - E_k}{\tilde{A}_2\, k_+} \\ 1 \end{pmatrix},$$

$$u_+ = \frac{\tilde{A}_2\, k}{\sqrt{\tilde{A}_2^2 k^2 + \left(Bk^2 - \varepsilon_g/2 - E_k\right)^2}} \begin{pmatrix} i\dfrac{\varepsilon_g/2 - Bk^2 + E_k}{\tilde{A}_2\, k_+} \\ 1 \end{pmatrix}. \tag{3.7}$$

The topological properties of the Bloch spinors (3.7) are characterized by the Berry field (curvature) (see Chap. 2, (2.43)). Direct calculation of the Berry field gives

$$F_{z\pm} = i\left(\left\langle \frac{\partial u_\pm}{\partial k_x} \middle| \frac{\partial u_\pm}{\partial k_y} \right\rangle - \left\langle \frac{\partial u_\pm}{\partial k_y} \middle| \frac{\partial u_\pm}{\partial k_x} \right\rangle\right) = \pm \frac{\tilde{A}_2^2(\varepsilon_g/2 + Bk^2)}{2E_k^3}, \tag{3.8}$$

and it is illustrated in Figs. 3.7 and 3.8. The valence band Berry curvature for the sufficiently strong exchange field (see Fig. 3.4b, c) is shown in Fig. 3.7.

The Berry field (3.8) is calculated within the $k - p$ model that is valid in the vicinity of the Γ-point of the BZ. Nevertheless, this model includes the main contribution which comes from k−regions of "singularities" related to the band inversion as shown in Fig. 3.7. At large k, outside the singularity region, the Berry field fades as $1/k^4$, allowing extrapolation (3.8) to the entire BZ. Then, an observable quantity, that is the Berry field integrated over BZ, can be calculated, extending the integration to infinite limits.

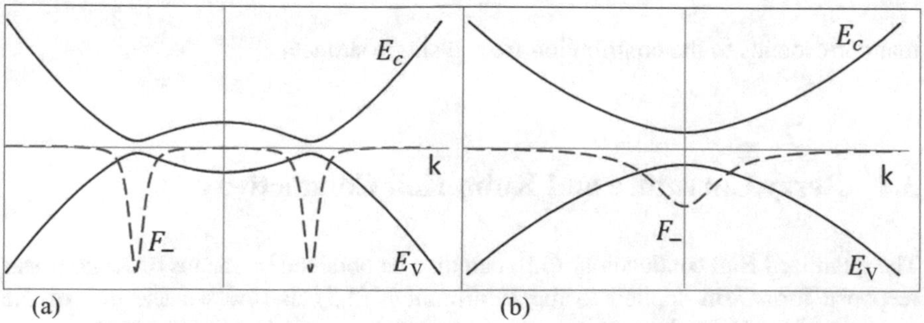

Fig. 3.7 The valence band Berry curvature (dashed) in an inverted spectrum, $m_z > \varepsilon_g/2$, $\varepsilon_g = 0.1$ eV. **(a)** $\tilde{A}_2 = 0.3$ eVÅ. **(b)** $\tilde{A}_2 = 3$ eVÅ. Different scales are used for the Berry field and energy bands

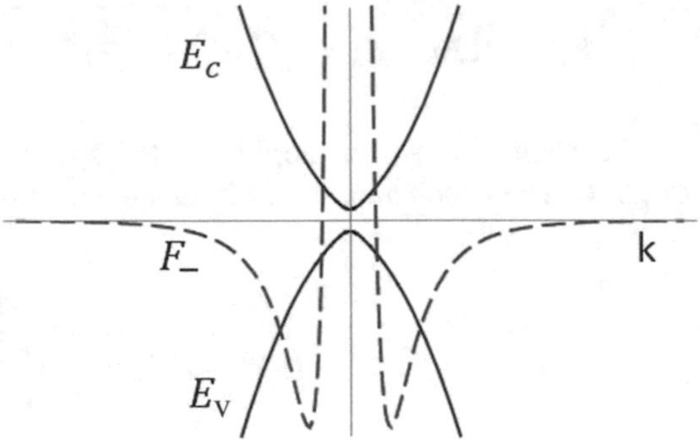

Fig. 3.8 Valence Berry curvature (dashed) in the direct spectrum, $m_z < \varepsilon_g/2$

If the exchange field is weak (see Fig. 3.4a), the Berry field changes sign in the BZ, as shown in Fig. 3.8. Below, we show that in that case, the Berry field integrated over the BZ gives zero.

The contribution to the Hall conductivity (see Chap. 2, (2.46)) from the filled valence band can be calculated as a Chern number by integrating the Berry field (3.8) as follows:

$$\sigma_{xy} = \frac{e^2}{2\pi h}\int_0^\infty kdk\int_0^{2\pi} d\varphi F_{z-}(k) = -\left[\text{sgn}(B) + \text{sgn}(\varepsilon_g)\right]\frac{e^2}{2h}. \tag{3.9}$$

The Chern number ($C_1 = \pm 1$) on the right hand side of (3.9) describes the quantized intrinsic QAHE: the Hall conductivity jumps from zero to the value of e^2/h which is insensitive to the parameters in (3.8) as far as they satisfy the condition $B\varepsilon_g > 0$. Sufficiently strong hybridization between two Dirac cones on opposite surfaces ($\Delta_S > |m_z|$) destroys the quantization of the Hall conductivity. In the opposite limit of independent surfaces ($B, \Delta_S \to 0$), calculations in (3.9) result in a half-integer quantization

$$\sigma_{xy} = -\text{sgn}(m_z)\frac{e^2}{2h} \tag{3.10}$$

that corresponds to the contribution from a single surface.

3.4 Berry Curvature and Kubo Hall Conductivity

The quantized Hall conductivity (3.9) can also be obtained by means of Kubo linear response formalism applied to the Hamiltonian (3.5). Below we use one of the versions of the Kubo formula as a response to a constant electric field [19]:

$$\sigma_{xy} = \frac{i}{\omega} Q_{xy}(\omega + i\delta), \quad \omega \to 0,$$

where

$$Q_{xy}(i\nu_m) = \frac{e^2}{\beta\hbar S} \sum_{\omega_n} \sum_{k_x k_y} Tr\left[\mathrm{v}_x G(\boldsymbol{k}, i\omega_n) \mathrm{v}_y G(\boldsymbol{k}, i\omega_n + i\nu_m) \right], \tag{3.11}$$

S is the sample area, $\beta = T^{-1}$, T is the temperature in energy units,

$$\mathrm{v}_x = \frac{\partial H}{\partial k_x} = \begin{pmatrix} -2Bk_x & i\tilde{A}_2 \\ -i\tilde{A}_2 & 2Bk_x \end{pmatrix}, \quad \mathrm{v}_y = \begin{pmatrix} -2Bk_y & \tilde{A}_2 \\ \tilde{A}_2 & 2Bk_y \end{pmatrix}, \tag{3.12}$$

and the Matsubara Green function,

$$G(\boldsymbol{k}, i\omega_n) = (i\omega_n - H)^{-1}$$
$$= \frac{1}{(i\omega_n + E_k)(i\omega_n - E_k)} \begin{pmatrix} \varepsilon_g/2 + i\omega_n - Bk^2 & i\tilde{A}_2(k_x - ik_y) \\ -i\tilde{A}_2(k_x + ik_y) & -\varepsilon_g/2 + i\omega_n + Bk^2 \end{pmatrix}. \tag{3.13}$$

Performing the trace operation, we obtain the following expression:

$$\frac{1}{S} \sum_{k_x k_y} Tr\left[\mathrm{v}_x G(\boldsymbol{k}, i\omega_n) \mathrm{v}_y G(\boldsymbol{k}, i\omega_n + i\nu_m) \right] = \int_0^\infty kdk \frac{-2i\tilde{A}_2^2\, i\nu_m\, (\varepsilon_g/2 + Bk^2)}{\left(E_k^2 - (i\omega_n)^2\right)\left(E_k^2 - (i\nu_m + i\omega_n)^2\right)}. \tag{3.14}$$

Assuming that the chemical potential is equal to zero, i.e., lies in the middle of the gap, the summation over fermion Matsubara frequencies can be performed using the relation

$$\frac{1}{\beta} \sum_{\omega_n} g(i\omega_n) = -\frac{1}{2} \sum_{\mathrm{p}} \mathrm{Res}\left\{ g(z_{\mathrm{p}})\, tanh\left[\frac{\beta z_{\mathrm{p}}}{2}\right] \right\}, \tag{3.15}$$

where the sum runs over four poles of the function $g(z) : z_{\mathrm{p}} = \pm E_k, \pm E_k - i\nu_m$. After summation in (3.15), we arrive at the Hall conductivity

$$\sigma_{xy} = -\frac{e^2}{h} \int_0^\infty kdk \frac{\tilde{A}_2^2(\varepsilon_g/2 + Bk^2)\, tanh[\beta E_k/2]}{2E_k^3}, \tag{3.16}$$

which at $T \to 0$ coincides with expression (3.9) derived directly from the Berry curvature.

The temperature factor in (3.16) can be written through Fermi distribution functions as $tanh[\beta E_k/2] = f(-\beta E_k) - f(\beta E_k)$. If the chemical potential μ is in the conduction (or valence) band, summation over Matsubara frequencies gives the population factor as $f(-\beta(E_k + \mu)) - f(\beta(E_k - \mu))$. At $T \rightarrow 0$, this factor cuts off the k-integral in (3.16) at the lower limit of k_F, reducing the range of integration from the whole Brillouin zone to its part. The resulting σ_{xy} depends on the Berry curvature and can be finite, describing the AHE in metals. However, in this case, the Hall conductivity is not a topological invariant. In other words, σ_{xy} is not quantized as it is when the integration goes over the closed surface of the entire Brillouin zone resulting in the Chern number (see Chap. 2).

3.4.1 Generic Two-Band Model and Skyrmion Topology

As the Hamiltonian (3.5) is a 2×2 matrix, it can be represented as a linear combination of Pauli matrices in generic form $h(\boldsymbol{k}) = \boldsymbol{d}(\boldsymbol{k})\boldsymbol{\sigma}$, where $\boldsymbol{d}(\boldsymbol{k})$ is the vector comprising matrix elements of the Hamiltonian $H_{\boldsymbol{k}}$. The Hamiltonian (3.5) describes low-energy electron excitations with wave vectors close to the Dirac point in the BZ (Γ-point in $Bi_2(TeSe)_3$ material system). A minimal tight-binding model that could describe the QAHE was proposed in [19]. It is defined in the entire BZ, where the vector $\boldsymbol{d}(\boldsymbol{k})$ is given by

$$d_x = A \sin k_x \ ; \quad d_y = -A \sin k_y; \quad d_z = M - 2B\left(2 - \cos k_x - \cos k_y\right), \quad (3.17)$$

where A, M, B are the material constants of spin–orbit coupling, gap parameter, and dispersion, respectively, $MB > 0$.

Calculating the Hall conductivity with the procedure given in the preceding section, one gets a generic result expressed in terms of $\boldsymbol{d}(\boldsymbol{k})$:

$$\sigma_{xy} = \frac{e^2}{4\pi h} \int_{\mathrm{BZ}} dk_x dk_y \left(\frac{\partial \hat{\boldsymbol{d}}}{\partial k_x} \times \frac{\partial \hat{\boldsymbol{d}}}{\partial k_y} \right) \cdot \hat{\boldsymbol{d}}, \quad (3.18)$$

where $\hat{\boldsymbol{d}} = \boldsymbol{d}/\mid \boldsymbol{d} \mid$ is the unit vector in the $\boldsymbol{d}$ direction.

While $\boldsymbol{k}$ runs over the BZ, the vector $\hat{\boldsymbol{d}}$ covers the surface of a unit sphere, and the integrand in (3.18) is the Jacobian that translates integration over the BZ torus into integration over the whole spherical surface. The integral is equal to $4\pi n$, that is the area the vector $\hat{\boldsymbol{d}}$ covers when $\boldsymbol{k}$ runs over the torus: the full sphere area of 4π multiplied by the number of times the vector $\hat{\boldsymbol{d}}$ covers the sphere, the winding number n. If the winding number is equal to one, the structure of vector $\hat{\boldsymbol{d}}(\boldsymbol{k})$ forms the skyrmion configuration in k-space as illustrated in Fig. 3.9 by depicting the $\hat{d}_z$-component as a function of $k_{x,\,y}$ in the entire BZ.

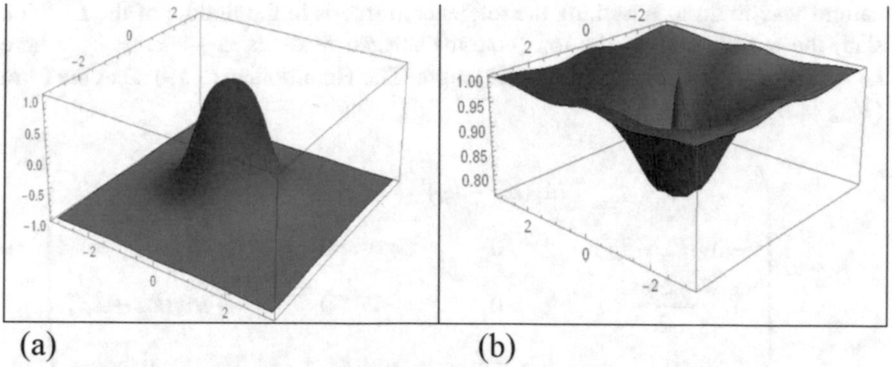

Fig. 3.9 Skyrmion in momentum space. Images depict $\hat{d}_z(k_x, k_y)$ only. (**a**) "Point-like" twist (-1 to 1) of the vector $\hat{\boldsymbol{d}}(\boldsymbol{k})$, $MB > 0$. (**b**) normal phase, $MB < 0$

As shown in Fig. 3.9a, vectors $\hat{\boldsymbol{d}}(\boldsymbol{k})$ at the BZ boundaries and at the center point in opposite directions. The structure of the skyrmion is topologically stable. That means the π-twist of $\hat{\boldsymbol{d}}(\boldsymbol{k})$ does not change if parameters A, M, B change continuously until the condition $MB > 0$ is violated and then the twist disappears, as shown in Fig. 3.9b.

Skyrmion objects in real space exist in magnetic materials, where the Dzyaloshinski-Moria exchange interaction makes the point-like twist of magnetization vector [20–24] energetically favorable.

3.5 Quantum Phase Transition Driven by a Parallel Magnetic Field

We start with the Hamiltonian of a TI film (3.1) written in the top-bottom representation (see Chap. 1, (1.48)):

$$H_{\mathrm{tb}} = \begin{pmatrix} m_z + \tilde{V}_{AS} & i\tilde{A}_2 k_- & -\dfrac{\Delta_S}{2} + Bk^2 & 0 \\ -i\tilde{A}_2 k_+ & -m_z + \tilde{V}_{AS} & 0 & -\dfrac{\Delta_S}{2} + Bk^2 \\ -\dfrac{\Delta_S}{2} + Bk^2 & 0 & m_z - \tilde{V}_{AS} & -i\tilde{A}_2 k_- \\ 0 & -\dfrac{\Delta_S}{2} + Bk^2 & i\tilde{A}_2 k_+ & -m_z - \tilde{V}_{AS} \end{pmatrix}. \tag{3.19}$$

Instead of proximity magnetization ($m_z = 0$), we apply an in-plane magnetic field B_y and take the Dirac limit ($Bk^2 \rightarrow 0$). In order to use the Peierls substitution $\boldsymbol{p} \rightarrow \boldsymbol{p} + e\boldsymbol{A}$, the vector potential in the Landau gauge, $A_x = -zB_y$ has to be specified on the top and bottom surfaces. The top-bottom representation provides a natural way to do it. Assuming the reference frame is in the middle of the L−thick slab, the substitution on the top (bottom) surface reads as $k_x \to k_x \pm k_B$, where $k_B = L/2l^2, l = \sqrt{\hbar/eB}$ is the magnetic length. The Hamiltonian (3.19) takes the form $(\tilde{V}_{AS} = 0)$:

$$H_{\mathrm{tb}} = \begin{pmatrix} 0 & i\tilde{A}_2(k_- - k_B) & -\frac{\Delta_S}{2} & 0 \\ -i\tilde{A}_2(k_+ - k_B) & 0 & 0 & -\frac{\Delta_S}{2} \\ -\frac{\Delta_S}{2} & 0 & 0 & -i\tilde{A}_2(k_- + k_B) \\ 0 & -\frac{\Delta_S}{2} & i\tilde{A}_2(k_+ + k_B) & 0 \end{pmatrix}. \tag{3.20}$$

The spectrum of the Hamiltonian (3.20) is given below

$$E = \pm\sqrt{\frac{\Delta_S^2}{4} + \tilde{A}_2^2(k^2 + k_B^2) \pm \tilde{A}_2\, k_B\sqrt{\left(\frac{\Delta_S^2}{4} + \tilde{A}_2^2 k_x^2\right)}}\,. \tag{3.21}$$

If surfaces are independent ($\Delta_S = 0$), the k_x-dispersion presents Dirac cones shifted $2k_B$ apart as shown in Fig. 3.10.

In thin films ($\Delta_S \neq 0$), the evolution of the spectrum (3.21) with Δ_S corresponds to the semimetal-semiconductor quantum phase transition as shown in Fig. 3.11.

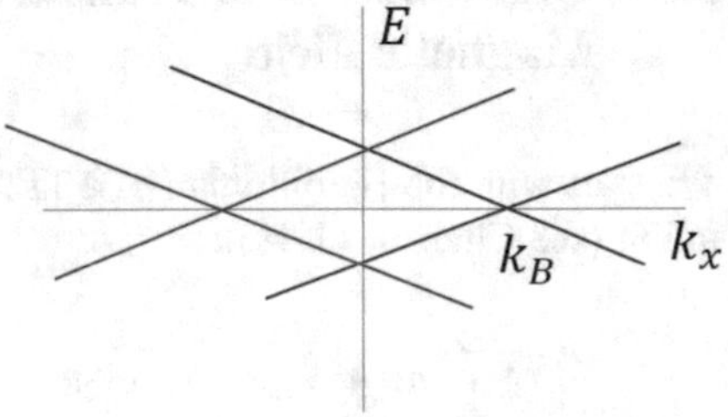

Fig. 3.10 Two sets of Dirac cones on opposite surfaces, $k_y = 0$, $\Delta_S = 0$

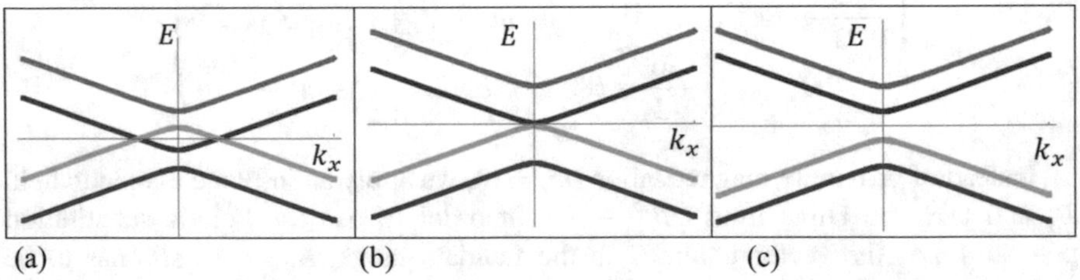

Fig. 3.11 Spectrum (3.21) at $k_y = 0$. **(a)** $0 < \Delta_S < 2\tilde{A}_2 k_B$, **(b)** $\Delta_S = 2\tilde{A}_2 k_B$, **(c)** $\Delta_S > 2\tilde{A}_2 k_B$

The dielectric phase shown in Fig. 3.11c can be identified by a nonzero diamagnetic response [25].

3.6 Thin Film in a Perpendicular Magnetic Field. Parity Anomaly

We consider the slab in a perpendicular magnetic field with the vector potential $A_x = yB_z$. The Hamiltonian (3.19) ($Bk^2 \ll \Delta_S$) takes the form:

$$H_{\text{tb}} = \begin{pmatrix} m_z + \tilde{V}_{AS} & i\text{v}\left(p_x + eA_x - ip_y\right) & -\frac{\Delta_S}{2} & 0 \\ -i\text{v}\left(p_x + eA_x + ip_y\right) & -m_z + \tilde{V}_{AS} & 0 & -\frac{\Delta_S}{2} \\ -\frac{\Delta_S}{2} & 0 & m_z - \tilde{V}_{AS} & -i\text{v}\left(p_x + eA_x - ip_y\right) \\ 0 & -\frac{\Delta_S}{2} & i\text{v}\left(p_x + eA_x + ip_y\right) & -m_z - \tilde{V}_{AS} \end{pmatrix}, \tag{3.22}$$

where $m_z = g\mu_B B_z/2$ is the Zeeman energy, $\text{v} = \tilde{A}_2/\hbar$. In order to find Landau levels we substitute momenta by ladder operators: $p_x + eA_x + ip_y = \sqrt{2e\hbar B_z}a^+$, $p_x + eA_x - ip_y = \sqrt{2e\hbar B_z}\, a$, $[aa^+] = 1$:

$$H_{\text{tb}} = \begin{pmatrix} m_z + \tilde{V}_{AS} & i\hbar\omega_c a & -\frac{\Delta_S}{2} & 0 \\ -i\hbar\omega_c a^+ & -m_z + \tilde{V}_{AS} & 0 & -\frac{\Delta_S}{2} \\ -\frac{\Delta_S}{2} & 0 & m_z - \tilde{V}_{AS} & -i\hbar\omega_c a \\ 0 & -\frac{\Delta_S}{2} & i\hbar\omega_c a^+ & -m_z - \tilde{V}_{AS} \end{pmatrix}, \tag{3.23}$$

where the cyclotron energy reads as $\hbar\omega_c = \tilde{A}_2\sqrt{2eB_z/\hbar}$. The operators a, a^+ are defined on the Hilbert space of orthonormalized oscillator functions: $a\varphi_n = \sqrt{n}\,\varphi_{n-1}$, $a^+\varphi_n = \sqrt{n+1}\,\varphi_{n+1}$, $n \in N = 0, 1, 2\ldots$ Choosing the wave function as a spinor composed of oscillator functions, $\Psi^T = \{f_1\,\varphi_{n-1}, f_2\,\varphi_n, f_3\,\varphi_{n-1}, f_4\,\varphi_n\}$, we change the representation, i.e., go to the Hamiltonian acting in the Hilbert space of coefficients f_i. Let's take as an example the first line of the Schrödinger equation $H_{\text{tb}}\Psi = E\Psi$:

$$-\frac{\Delta_S}{2} f_3\varphi_{n-1} + \left(m_z + \tilde{V}_{AS}\right) f_1\varphi_{n-1} + i\,\hbar\omega_c\, f_2 a\varphi_n = Ef_1\,\varphi_{n-1}. \tag{3.24}$$

Multiplying the equation by φ^*_{n-1} and integrating over a spatial variable of oscillator functions one gets

$$-\frac{\Delta_S}{2}f_3+(m_z+\tilde{V}_{AS})f_1+i\hbar\omega_c f_2\sqrt{n}=Ef_1. \tag{3.25}$$

Gathering three other lines, we get the final Hamiltonian in the form:

$$H=\begin{pmatrix} m_z+\tilde{V}_{AS} & i\hbar\omega_c\sqrt{n} & -\frac{\Delta_S}{2} & 0 \\ -i\hbar\omega_c\sqrt{n} & -m_z+\tilde{V}_{AS} & 0 & -\frac{\Delta_S}{2} \\ -\frac{\Delta_S}{2} & 0 & m_z-\tilde{V}_{AS} & -i\hbar\omega_c\sqrt{n} \\ 0 & -\frac{\Delta_S}{2} & i\hbar\omega_c\sqrt{n} & -m_z-\tilde{V}_{AS} \end{pmatrix}, \tag{3.26}$$

with corresponding eigenvalues

$$E_n=\pm\sqrt{\frac{\Delta_S^2}{4}+\left(m_z^2+n\hbar^2\omega_c^2+\tilde{V}_{AS}^2\pm 2\sqrt{\frac{\Delta_S^2}{4}m_z^2+\left(m_z^2+n\hbar^2\omega_c^2\right)\tilde{V}_{AS}^2}\right)}, \tag{3.27}$$

where the signs in front of the square root correspond to electron- or hole-like Landau levels. The four sublevels (3.27) present the Landau spectrum split by the simultaneous presence of the top-bottom coupling, the Zeeman field, and the vertical bias. If $\Delta_S,\ m_z,\ \tilde{V}_{AS}\to 0$, the spectrum presents the Landau levels of a Dirac cone, $E_n=\pm\hbar\mid\omega_c\mid\sqrt{n}$, where $\pm$ corresponds to the conduction and valence states.

The Landau level with $n=0$ should be treated separately as the spinor $\Psi^T(n=0)=\{0,f_2\,\varphi_n,0,f_4\,\varphi_n\}$, being substituted into equation $H_{\text{tb}}(n=0)\Psi=E\Psi$, gives only two equations for coefficients $f_{2,\,4}$:

$$\begin{aligned} -\frac{\Delta_S}{2}f_4+(-m_z+\tilde{V}_{AS})f_2&=Ef_2,\\ -\frac{\Delta_S}{2}f_2+(-m_z-\tilde{V}_{AS})f_4&=Ef_4, \end{aligned} \tag{3.28}$$

Instead of four levels at each $n\neq 0$ in (3.27), the (3.28) result in two levels of the same spin orientation: $E_0=-m_z\pm\sqrt{\Delta_S^2/4+\tilde{V}_{AS}^2}$. The top-bottom hybridization and vertical bias split the zero level in two. If the splitting is small the sublevels preserve their character being both hole-like ($m_z>0$) or electron-like ($m_z<0$). However, if the splitting is larger than the magnetic gap, one of the sublevels changes the band to which the level belongs. In the inversion symmetric case $(\tilde{V}=0)$, it was shown in [26].

If the surfaces are independent, one may consider a single Dirac cone. The Zeeman mass term, $m_z\neq 0$, may originate either from the magnetic field or from

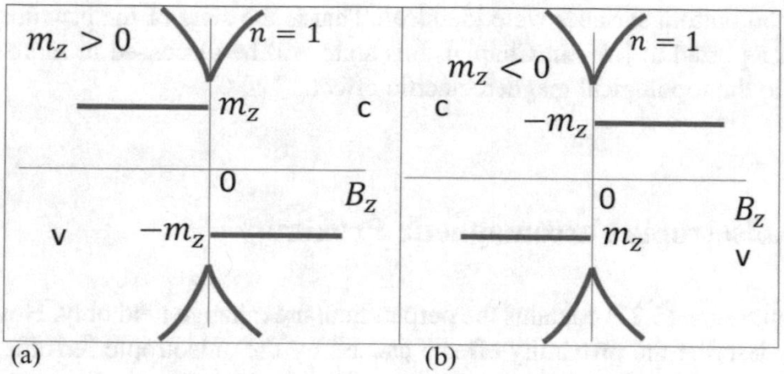

Fig. 3.12 Landau levels vs. the magnetic field B_z. The zero Landau level manifests a parity anomaly: the level belongs to the c(v) band depending on the sign of m_z. **(a)** $m_z > 0$. **(b)** $m_z < 0$

the proximity exchange field creating the energy gap in the Dirac spectrum. If the mass term is present, all $n \neq 0$ levels form the conduction and valence sets which are even with respect to the magnetic field, $E_n(B_z) = E_n(-B_z)$, as shown in Fig. 3.12.

The zero Landau level E_0 is electron- or hole-like depending on the sign of the gap m_z. As illustrated in Fig. 3.12, the zero level position is an odd function of the magnetic field B_z. That is the trace of a parity anomaly of the Dirac solid state system [27, 28]. When $m_z \to 0$, the zero mode is shared between conduction and valence bands.

If two surfaces are independent ($\Delta_S = 0$) and the Fermi level is in between the conduction Landau levels, the Dirac-fermion Hall conductivity on both surfaces together is given as

$$\sigma_{xy} = \frac{2e^2}{h}\left(n + \frac{1}{2}\right). \tag{3.29}$$

Factor 2 in (3.29) reflects top-bottom degeneracy, in other words it describes the contributions from two equal sets of Dirac-type Landau levels on opposite surfaces, so the odd filling factor of $\nu = \pm(2n+1)$ could be seen experimentally. However, the experimental samples are grown on substrates that make the two surfaces electrically biased with respect to each other. The bias breaks the spatial inversion symmetry, lifting the top-bottom degeneracy in the Dirac cones, so that the modified spectrum becomes

$$E_n = \pm\tilde{V}_{AS} \pm \sqrt{m_z^2 + n\,\hbar^2\omega_c^2}, \quad \Delta_S = 0. \tag{3.30}$$

Two sets of Dirac Landau levels are shifted in energy relative to each other by the vertical bias, so the QHE experiment would give contributions to the Hall conductivity from the top and bottom surfaces as $\sigma_{xy} = \frac{e^2}{h}(n + n' + 1)$. If the Fermi level is placed in the gap between two Dirac points on opposite surfaces, the observable Hall conductivity turns to zero ($n = 0, n' = -1$). This would not otherwise be possible if the top and bottom surfaces were identical. That is the state of the pseudospin Hall insulator reported in [29]. In Chap. 4, this state will be discussed in more detail in relation to the topological magnetoelectric effect.

3.7 Anisotropic Ferromagnetic Proximity

The Hamiltonian (3.22) contains the perpendicular exchange field only. However, in order to describe the proximity effects caused by the anisotropic ferromagnet, we have to account for all components of the exchange field. The exchange field terms enter an effective surface Hamiltonian as matrix elements of the interaction $J\boldsymbol{\sigma}\boldsymbol{M}$, where J is the exchange constant, $\boldsymbol{M}$ is the magnetization in the ferromagnetic layer, and $\boldsymbol{\sigma}$ is the operator of electron spin. The matrix elements are calculated with the surface wave functions (1.21) and (1.25) and enter the nondiagonal terms of the Hamiltonian (3.19) as follows:

$$H_{\mathrm{tb}} = \begin{pmatrix} m_z + \tilde{V}_{AS} & i\tilde{A}_2 k_- + im_x + m_y & -\frac{\Delta_S}{2} + Bk^2 & 0 \\ -i\tilde{A}_2 k_+ - im_x + m_y & -m_z + \tilde{V}_{AS} & 0 & -\frac{\Delta_S}{2} + Bk^2 \\ -\frac{\Delta_S}{2} + Bk^2 & 0 & m_z - \tilde{V}_{AS} & -i\tilde{A}_2 k_- - im_x - m_y \\ 0 & -\frac{\Delta_S}{2} + Bk^2 & i\tilde{A}_2 k_+ + im_x - m_y & -m_z - \tilde{V}_{AS} \end{pmatrix}, \tag{3.31}$$

where $m_i = JM_i$ are the exchange field corrections to the electron energy.

The ferromagnetic exchange field extends approximately 2 nm deep into a TI [12], justifying the assumption of nonhomogeneous field distribution across the film thickness. In thin films, we keep the top-bottom tunneling terms and assume that the exchange corrections are homogenous from the top to the bottom surfaces. In a thick film, the magnitudes of the exchange field might be different at the top and bottom surfaces and one may consider the exchange field to be localized near the TI/FM interface only (let's say, the top surface of the TI).

If we deal with the Dirac model in (3.31) ($Bk^2 \ll \Delta_S/2$) or with the independent surfaces ($B = \Delta_S = 0$), the in-plane components of the exchange field enter the Hamiltonian like the vector potential: $k_{x,\,y} \to k_{x,\,y} + \mathrm{const} * m_{x,\,y}$. In the homogeneous proximity field, the electron spectrum of TI found from (3.31) is expressed as

$$E_{\mathrm{c}\uparrow\downarrow,\mathrm{v}\uparrow\downarrow} = \pm\left[\left(\tfrac{\Delta_S}{2} - Bk^2\right)^2 + \left(\tilde{A}_2\boldsymbol{k} + \boldsymbol{m}\right)^2 + \tilde{V}_{AS}^2 \pm \sqrt{\left(\tfrac{\Delta_S}{2} - Bk^2\right)^2 m_z^2 + \tilde{V}_{AS}^2\left(\tilde{A}_2\boldsymbol{k} + \boldsymbol{m}\right)^2}\right]^{1/2},$$

$$\boldsymbol{m} = \left(m_x, m_y, m_z\right); \quad \boldsymbol{k} = \left(k_x, k_y, 0\right). \tag{3.32}$$

If one neglects the vertical bias and accounts for the out-of-plane anisotropy only, $(m_{x,y} = 0, \tilde{V}_{AS} = 0)$, the spectrum (3.32) coincides with (3.3). The effective masses and energy gaps in (3.32) depend on magnetic anisotropy in the FM layer. The anisotropy itself can be controlled by the thickness of ferromagnet deposited onto a TI surface [30]. Switching electron dispersion by controlling the magnetic anisotropy is illustrated in Fig. 3.13 which shows the spectrum when the magnetic anisotropy varies. The parameters used in numerical examples are $\Delta_S = 0.1\,\text{eV},\ B = 30\,\text{eV}\dot{\text{A}}^2,\ \tilde{V}_{AS} = 0.1\ \text{eV},\ \tilde{A}_2 = 4.1\ \text{eV}\dot{\text{A}},\ k_y = m_y = 0.$

The lateral proximity field, acting equally on the top and bottom surfaces, equally shifts both Dirac cones in k-space while the vertical bias separates the cones on an energy scale (Fig. 3.13a). The out-of-plane anisotropy component opens gaps in the Dirac points (Fig. 3.13b). The dispersion in the presence of perpendicular anisotropy only is shown in Fig. 3.13c.

In thick films, one can neglect the top-bottom mixing terms and, because of the short range of exchange interaction, one can assume that the proximity field is relevant only on the top TI surface adjacent to the FM layer. The sequence of basic functions in the representation (3.31) implies that two 2 × 2 blocks along a main diagonal correspond to the bottom and top electrons, respectively. The corresponding Hamiltonian describes independent surfaces and follows from (3.31):

$$H_{\text{tb}} = \begin{pmatrix} \tilde{V}_{AS} & i\tilde{A}_2(k_x - ik_y) & 0 & 0 \\ -i\tilde{A}_2(k_x + ik_y) & \tilde{V}_{AS} & 0 & 0 \\ 0 & 0 & m_z - \tilde{V}_{AS} & -i\tilde{A}_2(k_x - ik_y) - im_x - m_y \\ 0 & 0 & i\tilde{A}_2(k_x + ik_y) + im_x - m_y & -m_z - \tilde{V}_{AS} \end{pmatrix}, \tag{3.33}$$

The block-diagonal Hamiltonian (3.33) describes branches that belong to a particular surface:

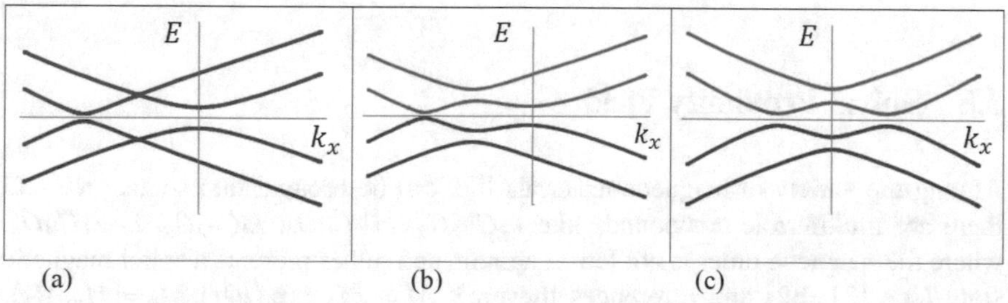

Fig. 3.13 Surface electron dispersion modified by a homogeneous proximity field. Exchange field direction: (**a**) in-plane exchange field, $m_z = 0$, $m_x = 0.1$ eV. (**b**) tilted exchange field, $m_x = 0.1$ eV, $m_z = 0.03$ eV; (**c**) out-of-plane exchange field, $m_x = 0$, $m_z = 0.03$ eV

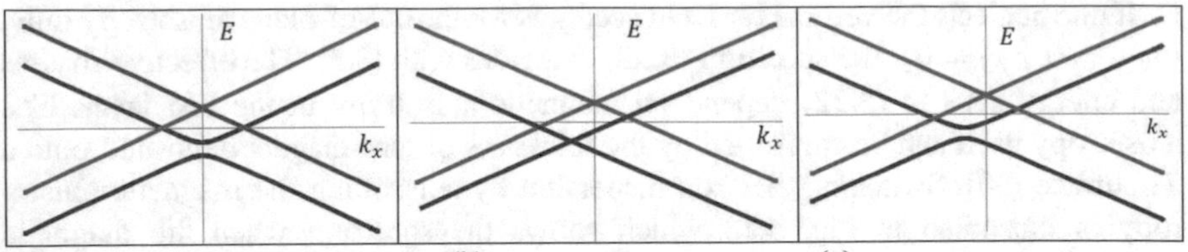

Fig. 3.14 Surface electrons in an inhomogeneous proximity field, $\tilde{V}_{AS} = 0.1$ eV. (**a**) out-of-plane anisotropy, $m_x = 0$, $m_z = 0.03$ eV. (**b**) in-plane anisotropy, $m_x = 0.1$ eV, $m_z = 0$. (**c**) tilted magnetization, $m_x = 0.1$ eV, $m_z = 0.03$ eV

$$E_{\rm t} = \pm\sqrt{\tilde{A}_2^2 k^2 + m^2 + 2\tilde{A}_2 \boldsymbol{k}\boldsymbol{m}_{\|}} - \tilde{V}_{AS},$$

$$E_{\rm b} = \pm\tilde{A}_2 \boldsymbol{k} + \tilde{V}_{AS}, \tag{3.34}$$

(a) The proximity field inhomogeneous across the film thickness acts on the top surface only and allows manipulating its electron spectrum, not affecting the opposite surface where the Dirac cone remains unchanged. The corresponding spectrum is illustrated in Fig. 3.14.

The perpendicular-to-plane magnetization induces the gap in the top Dirac cone, leaving the bottom one unchanged (Fig. 3.14a). In-plane anisotropy shifts the top Dirac cone in k-space relative to the bottom cone at $k = 0$ (Fig. 3.14b). The tilted magnetic moment makes the dispersion shifted in k_x and asymmetrically gapped (Fig. 3.14c).

The proximity-induced ferromagnetism can be identified by the AHE. If the surface spectrum is gapped and the Fermi energy is in the gap, a half-quantized QAHE is observed, where the Hall conductivity $\sigma_{xy} = e^2/h$ is the sum of contributions from both surfaces. In an experimental setup with a thick TI film where only one surface is gapped (see Fig. 3.14), the contribution to QAHE might result in $\sigma_{xy} = e^2/2h$. However, the Ohmic conductivity σ_{xx} on the gapless surface could shunt the conduction and destroy the QAHE.

3.8 Spiral Proximity Field

Among the variety of magnetic materials that can be brought into contact with TI there are multiferroic compounds like $LiCu_2O_2$, $LiVCuO_4$, $LiCu_2O_2$, Li_2ZrCuO_4, where the magnetic order is not ferromagnetic and rather presents a spiral magnetic state (see [31, 32] and references therein): $\boldsymbol{M} = \boldsymbol{M_0} \exp\left(iQr\right), \boldsymbol{M_0} = \left(M_x, M_y\right)$, $Q = \frac{2\pi}{L}, L$ is the period of the spiral. Half-period spiral ordering is shown in Fig. 3.15.

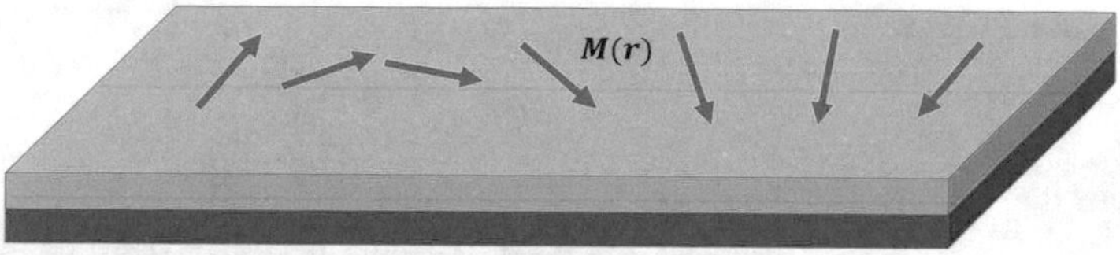

Fig. 3.15 Spiral magnetic layer in contact with a TI film

Under the proximity field shown in Fig. 3.15, the Dirac points remain gapless. However, gaps open at the boundary of the new 1D Brillouin zone at $k = \pm Q/2$ [33].

3.9 Lateral Heterostructures and Domain Walls

The exchange field direction and the vertical bias drive the surface electron spectrum through topological phase transitions. This control has been observed experimentally and may be used in spintronic device applications of topological insulators [30, 34]. Placing the ferromagnetic ribbon in contact with TI allows the lateral electron transport to be controlled, creating the inhomogeneous gap distribution along the direction parallel to the surface. An example of anomalous magnetoresistance induced by the proximity exchange field and studied in [35] is illustrated in Fig. 3.16.

The vertical component of the magnetization opens an energy gap in region 2, creating a barrier for tunneling electrons, and making the magnetoresistance dependent on the angles θ and φ. The ohmic conductance is suppressed at angles $\theta = 0, \pi$ and reaches the maximum at $\theta = \pi/2$ with the magnitude dependent on the in-plane angle φ.

Lateral heterostructure like that shown in Fig. 3.16 may consist of two FM layers that forms the double-barrier structure. If the magnetic moment is fixed in one layer and the other layer has the free magnetic moment, the structure is estimated to have the giant magnetoresistance and may serve as a TI-based magnetoresistive valve [36].

The magnetic proximity setting in Fig. 3.16 at $\theta \neq \pi/2$ affects the inverse spin-galvanic effect (see (1.57) Chap. 1) since electrons tunnel through the proximity-induced energy gap and their spin polarization depends on the transparency of the barrier. In this case, the spin polarization can be expressed through the tunneling conductance as follows [37, 38]:

$$\frac{m_x}{\mu_B} = \frac{eE_F V}{2\pi^2 \tilde{A}_2^2} \int_{-\pi/2}^{\pi/2} T(E_\mathrm{F}, \alpha) \sin \alpha d\alpha,$$

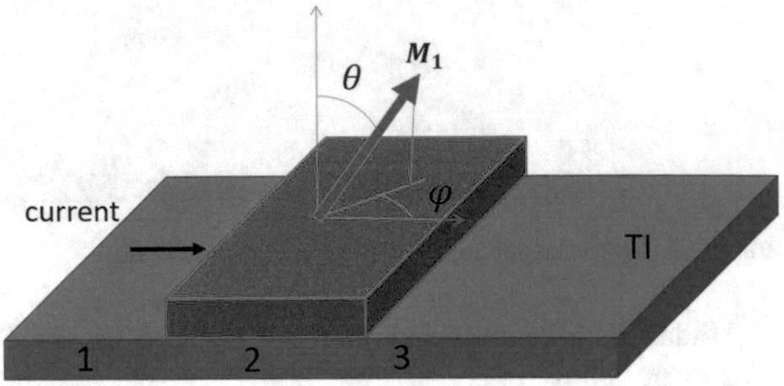

Fig. 3.16 Lateral barrier structure

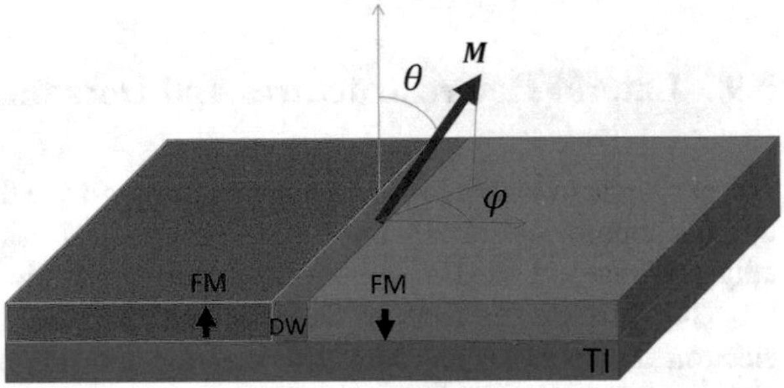

Fig. 3.17 Domain wall (shown in green). The energy gap is zero along the DW

$$\frac{m_y}{\mu_B} = -\frac{eE_F V}{2\pi^2 \tilde{A}_2^2} \int_{-\pi/2}^{\pi/2} T(E_F, \alpha) \cos \alpha \, d\alpha, \tag{3.35}$$

where $T(E_F, \alpha)$ is the transparency of the region 2 shown in Fig. 3.16, V is the voltage across that region, and α is the angle of incidence of Fermi electrons.

If the ferromagnetic order in FM breaks into two domains ordered in opposite directions ($\pm m_z$), as shown in Fig. 3.17, each domain contributes to the QAHE as $\sigma_{xy} = \pm \frac{e^2}{2h}$ (see (3.10)). The difference $\sigma_{xy} = e^2/h$ indicates the Chern number of 1 that corresponds to one chiral current mode flowing along the gapless domain wall (DW), where $m_z = 0$.

The energy spectrum of the edge mode depends on the type of DW. In Neel DW, the magnetic moment θ rotates in the plane where $\varphi = 0, \pi$, while in Bloch DW, the rotation is in the perpendicular plane, $\varphi = \frac{\pi}{2}, 3\pi/2$. The energy spectrum of the edge mode, calculated in [39], depends on φ and on the width of the domain wall. The stability of the domain wall itself depends on the filling of the edge channel. This creates a tool for the electrically controlled edge mode through the voltage-dependent Fermi level and allows the system to be used in device applications.

3.10 Proximity Coupling and Torque

Manipulation of the magnetic moment of a ferromagnetic materials (FM) by controlling the electrical current in an adjacent non-magnetic media is at the origin of various applications in memory devices. This phenomenon is based on the exchange interaction Δ_{ex} between the nonequilibrium electron magnetization $\boldsymbol{m}$ and magnetic moment $\boldsymbol{M}$ in a FM material. If $\boldsymbol{m}$ and $\boldsymbol{M}$ are not collinear, electron magnetic moment exerts spin-transfer torque (STT) on $\boldsymbol{M}$ [40, 41]. In a magnetic tunneling junction (MTJ) setting, the electron magnetization comes into a FM by the spin current injected through the interface when the vertical current flows through the MTJ. One more source of nonequilibrium spin polarization at the non-magnetic/FM interface originates from the spin–orbit interaction in a non-magnetic media. In this case, the torque is caused by proximity exchange coupling through the interface and is called spin–orbit torque (SOT) [42–44]. The SOT, in turn, has two sources of origin: the inverse spin-galvanic effect (see Chap. 1) and spin Hall effect. Both effects generate electron spin polarization [45]. Both mechanisms are present in topological insulators.

The torque acting on $\boldsymbol{M}$, $\boldsymbol{\tau} = \Delta_{ex}\boldsymbol{M} \times \boldsymbol{m}$, is not zero as long as the nonequilibrium $\boldsymbol{m}$ is not collinear to $\boldsymbol{M}$. Phenomenologically, the torque can be represented as a sum of in-plane and transverse components: $\boldsymbol{\tau} = \boldsymbol{\tau}_{\parallel} + \boldsymbol{\tau}_{\perp}$, where $\boldsymbol{\tau}_{\parallel} = a\,\Delta_{ex}\boldsymbol{M} \times (\boldsymbol{M} \times \boldsymbol{m})$, $\boldsymbol{\tau}_{\perp} = b\,\Delta_{ex}\boldsymbol{M} \times \boldsymbol{m}$, where a and b are the coefficients which should be determined from a microscopic transport theory. The components $\boldsymbol{\tau}_{\parallel}$ and $\boldsymbol{\tau}_{\perp}$ are often called the spin-transfer or anti-damping term and the field-like term, respectively. They are illustrated in Fig. 3.18.

As shown in Fig. 3.18b, the in-plane torque rotates $\boldsymbol{M}$ toward the direction of $\boldsymbol{m}$, while the transverse torque component results in the precession of $\boldsymbol{M}$ about the $\boldsymbol{m}$ direction. Nonlinear magnetic dynamics in this system are described by the Landau-Lifshits-Gilbert equation and are manifested in the ferromagnetic resonance [46], the current-induced $\boldsymbol{M}$-switching [14, 47–50] and temporal $\boldsymbol{M}$-oscillations [38, 51, 52].

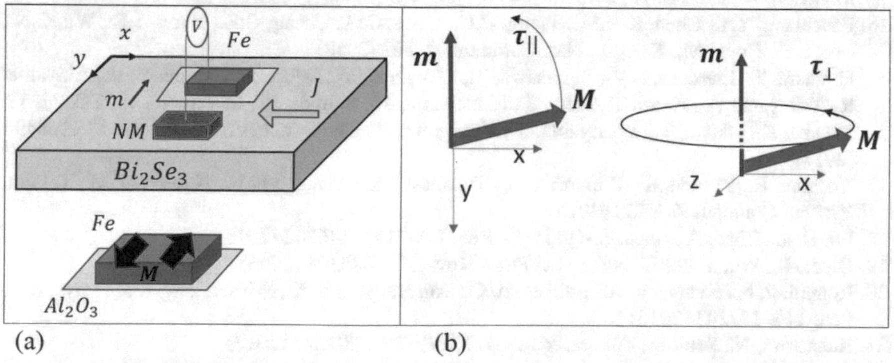

Fig. 3.18 (**a**) Torque (black arrows) induced by current-induced electron spin polarization at the TI surface exerted on the magnetic moment in a ferromagnetic layer. (**b**) Torque components rotate the FM magnetic moment

If the surface electron spectrum is gapped, the diagonal current vanishes and the inverse spin-galvanic effect is possible through the topological contribution caused by the quantum Hall effect, where the Hall current induces magnetization along the electric field [3]. This topological magnetoelectric effect will be discussed in Chap. 4.

References

1. Yu, R., Zhang, W., Zhang, H.J., Zhang, S.C., Dai, X., Fang, Z.: Science. **329**, 61 (2010)
2. Essin, A.M., Moore, J.E., Vanderbilt, D.: Phys. Rev. Lett. **102**, 146805 (2009)
3. Garate, I., Franz, M.: Phys. Rev. Lett. **104**, 146802 (2010)
4. Zhang, Y., He, K., Chang, C.Z., Song, C.L., Wang, L.L., Chen, X., Jia, J.F., Fang, Z., Dai, X., Shan, W.Y., Shen, S.Q., Niu, Q., Qi, X.L., Zhang, S.C., Ma, X.C., Xue, Q.K.: Nat. Phys. **6**, 584 (2010)
5. Chen, Y.L., Chu, J.H., Analitis, J.G., Liu, Z.K., Igarashi, K., Kuo, H.H., Qi, X.L., Mo, S.K.: Science. **329**, 659 (2010)
6. Chang, C.Z., Zhang, J., Feng, X., Shen, J., Zhang, Z., Guo, M., Li, K., Ou, Y., Wei, P., Wang, L.L., Ji, Z.Q., Feng, Y., Ji, S., Chen, X., Jia, J., Dai, X., Fang, Z., Zhang, S.C., He, K., Wang, Y., Lu, L., Ma, X.C., Xue, Q.K.: Science. **340**, 167 (2013)
7. Checkelsky, J.G., Yoshimi, R., Tsukazaki, A., Takahashi, K.S., Kozuka, Y., Falson, J., Kawasaki, M., Tokura, Y.: Nat. Phys. **10**, 731 (2014)
8. Kou, X., Guo, S., Fan, Y., Pan, L., Lang, M., Jiang, Y., Shao, Q., Nie, T., Murata, K., Tang, J., Wang, Y., He, L., Lee, T., Lee, W., Wang, K.L.: Phys. Rev. Lett. **113**, 137201 (2014)
9. Chang, C.-Z., Zhao, W., Kim, D.Y., Zhang, H.J., Assaf, B.A., Heiman, D., Zhang, S.C., Liu, C. X., Chan, M.H.W., Moodera, J.S.: Nat. Mater. **14**, 473 (2015)
10. Wei, P., Katmis, F., Assaf, B.A., Steinberg, H., Jarillo-Herrero, P., Heiman, D., Moodera, J.S.: Phys. Rev. Lett. **110**, 186807 (2013)
11. Jiang, Z., Chang, C.Z., Tang, C., Zheng, J.G., Moodera, J.S., Shi, J.: AIP Adv. **6**, 055809 (2016)
12. Katmis, F., Lauter, V., Nogueira, F.S., Assaf, B.A., Jamer, M.E., Wei, P., Satpati, B., Freeland, J.W., Eremin, I., Heiman, D., Jarillo-Herrero, P., Moodera, J.S.: Nature. **533**, 513 (2016)
13. Ralph, D.C., Stiles, M.D.: J. Magn. Magn. Mater. **320**, 1190 (2008)
14. Mellnik, R., Lee, J.S., Richardella, A., Grab, J.L., Mintun, P.J., Fischer, M.H., Vaezi, A., Manchon, A., Kim, E.-A., Samarth, N., Ralph, D.C.: Nature. **511**, 449 (2014)
15. Fanchiang, Y.T., Chen, K.H.M., Tseng, C.C., Chen, C.C., Cheng, C.K., Yang, S.R., Wu, C.N., Lee, S.F., Hong, M., Kwo, J.: Nat. Commun. **9**, 223 (2018)
16. Hirahara, T., Eremeev, S.V., Shirasawa, T., Okuyama, Y., Kubo, T., Nakanishi, R., Akiyama, R., Takayama, A., Hajiri, T., Ideta, S., Matsunami, M., Sumida, K., Miyamoto, K., Takagi, Y., Tanaka, K., Okuda, T., Yokoyama, T., Hasegawa, S., Chulkov, E.V.: Nano Lett. **17**(6), 3493 (2017)
17. Yoshimi, R., Yasuda, K., Tsukazaki, A., Takahashi, K.S., Nagaosa, N., Kawasaki, M., Tokura, Y.: Nat. Commun. **6**, 8530 (2015)
18. Lu, H.-Z., Zhao, A., Shen, S.-Q.: Phys. Rev. Lett. **111**, 146802 (2013)
19. Qi, X.-L., Wu, Y.-S., Zhang, S.-C.: Phys. Rev. **B74**, 085308 (2006)
20. Romming, N., Kubetzka, A., Hanneken, C., von Bergmann, K., Wiesendanger, R.: Phys. Rev. Lett. **114**, 177203 (2015)
21. Kanazawa, N., Seki, S., Tokura, Y.: Adv. Mater. **29**, 1603227 (2017)
22. Yu, G., Jenkins, A., Ma, X., Razavi, S.A., He, C., Yin, G., Shao, Q., He, Q.l., Wu, H., Li, W., Jiang, W., Han, X., Li, X., Bleszynski-Jayich, A.C., Amiri, P.K., Wang, K.L.: Nano Lett. **18**, 980 (2017)
23. Zhang, S., Kronast, F., van der Laan, G., Hesjedal, T.: Nano Lett. **18**, 1057 (2018)

24. Romming, N., Hanneken, C., Menzel, M., Bickel, J.E., Wolter, B., von Bergmann, K., Kubetzka, A., Wiesendanger, R.: Science. **341**, (2013)
25. Zyuzin, A.A., Hook, M.D., Burkov, A.A.: Phys. Rev. **B83**, 245428 (2011)
26. Zyuzin, A.A., Burkov, A.A.: Phys. Rev. **B83**, 195413 (2011)
27. Jackiw, R.: Phys. Rev. **D29**, 2375 (1984)
28. Haldane, F.D.M.: Phys. Rev. Lett. **61**, 2015 (1988)
29. Yoshimi, R., Tsukazaki, A., Kozuka, Y., Falson, J., Takahashi, K.S., Checkelsky, J.G., Nagaosa, N., Kawasaki, M., Tokura, Y.: Nat. Commun. **6**, 6627 (2015)
30. Honma, K., Sato, T., Souma, S., Sugawara, K., Tanaka, Y., Takahashi, T.: Phys. Rev. Lett. **115**, 266401 (2015)
31. Kobayashi, Y., Sato, K., Yasui, Y., Moyoshi, T., Sato, M., Kakurai, K.: J. Physical Soc. Japan. **78**, 084720 (2009)
32. Tokura, Y., Seki, S.: Adv. Mater. **22**, 1554 (2010)
33. Stagraczynski, S., Chotorlishvili, L., Dugaev, V.K., Jia, C.-L., Ernst, A., Komnik, A., Berakdar, J.: Phys. Rev. **B94**, 174436 (2016)
34. Wray, L.A., Xu, S.Y., Xia, Y., Hsieh, D., Fedorov, A.V., Hor, Y.S., Cava, R.J., Bansil, A., Lin, H., Hasan, M.Z.: Nat. Phys. **7**, 33 (2011)
35. Yokoyama, T., Tanaka, Y., Nagaosa, N.: Phys. Rev. **B81**, 121401 (2010)
36. Wu, Z., Peeters, F.M., Chang, K.: Phys. Rev. **B82**, 115211 (2010)
37. Büttiker, M., Imry, Y., Landauer, R., Pinhas, S.: Phys. Rev. **B31**, 6207 (1985)
38. Yokoyama, T.: Phys. Rev. **B84**, 113407 (2011)
39. Wakatsuki, R., Ezawa, M., Nagaosa, N.: Sci. Rep. **5**, 13638 (2015)
40. Slonczewski, J.C.: J. Magn. Magn. Mater. **159**, L1 (1996)
41. Berger, L.: Phys. Rev. **B54**, 9353 (1996)
42. Chernyshov, M.O., Liu, X., Furdyna, J.K., Lyanda-Geller, Y., Rokhinson, L.P.: Nat. Phys. **5**, 656 (2009)
43. Miron, M., Gaudin, G., Auffret, S., Rodmacq, B., Schuhl, A., Pizzini, S., Vogel, J., Gambardella, P.: Nat. Mater. **9**, 230 (2010)
44. Ramaswamy, R., Lee, J.M., Cai, K., Yang, H.: Appl. Phys. Lett. **5**, 031107 (2018)
45. Manchon, I. Miron, M., Jungwirth, T., Sinova, J., Zelezny, J., Thiaville, A., Garello, K., Gambardella, P.: Current-induced spin-orbit torques in ferromagnetic and antiferromagnetic systems, arXiv:1801.09636v1 [cond-mat.mes-hall]. Rev. Mod. Phys. (2018)
46. Yokoyama, T., Zang, J., Nagaosa, N.: Phys. Rev. **B81**, 241410(R) (2010)
47. Fischer, M.H., Vaezi, A., Manchon, A., Kim, E.A.: Phys. Rev. **B93**, 125303 (2016)
48. Ghosh, S., Manchon, A.: Phys. Rev. **B97**, 134402 (2018)
49. Fan, Y., Upadhyaya, P., Kou, X., Lang, M., Takei, S., Wang, Z., Tang, J., He, L., Chang, L.-T., Montazeri, M., Yu, G., Jiang, W., Nie, T., Schwartz, R.N., Tserkovnyak, Y., Wang, K.L.: Nat. Mater. **13**, 699 (2014)
50. Ndiaye, P.B., Akosa, C.A., Fischer, M.H., Vaezi, A., Kim, E.-A., Manchon, A.: Phys. Rev. **B96**, 014408 (2017)
51. Kiselev, S.I., Sankey, J.C., Krivorotov, I.N., Emley, N.C., Schoelkopf, R.J., Buhrman, R.A., Ralph, D.C.: Nature. **425**, 380 (2003)
52. Semenov, Y.G., Duan, X., Kim, K.W.: Phys. Rev. **B89**, 201405(R) (2014)

Chapter 4
Topological Magnetoelectric Effect

The electrodynamics of a media is described by the Maxwell equations which relate physical fields, current and charge density [1]. In four-dimensional Minkowski space $x^i = (ct, \boldsymbol{r})$ ($i = 0, 1, 2, 3, c$ is the speed of light in vacuum), with the metric tensor $g_{ij} =$ diag $\{1, -1, -1, -1, \}$, the inner product is defined as $X_i X^i = X^2$, where $X_i = g_{ij} X^j$ and over repeated indices, summation is implied. The potentials are expressed as components of the four-vector $A^i = (\varphi/c, \boldsymbol{A})$, $A_i = (\varphi/c, -\boldsymbol{A})$. The gauge-invariant electromagnetic field tensor is defined as

$$F_{ik} = \partial_i A_k - \partial_k A_i, \;\; \partial_i \equiv \frac{\partial}{\partial x^i}. \tag{4.1}$$

Differentiating the potential A_i we obtain the electromagnetic tensor expressed through the physical fields

$$F_{ik} = \begin{pmatrix} 0 & E_x/c & E_y/c & E_z/c \\ -E_x/c & 0 & -B_z & B_y \\ -E_y/c & B_y & 0 & -B_x \\ -E_z/c & -B_y & B_x & 0 \end{pmatrix}, \tag{4.2}$$

where $\boldsymbol{E} = -\partial \boldsymbol{A}/\partial t - \nabla\varphi$, $\boldsymbol{B} = \nabla \times \boldsymbol{A}$ are the electric field and magnetic flux density, respectively.

It can be checked by direct calculations in (4.1) that

$$\epsilon^{iklm} \partial_k F_{lm} = 0, \; i = 0{,}1{,}2{,}3, \tag{4.3}$$

where ϵ^{iklm} is the Levi-Civita symbol, which is equal to +1 (-1) if an even (odd) number of pairwise permutations bring indexes to the sequence 0,1,2,3. From (4.2) and (4.3) we find the first pair of Maxwell equations in vector form:

V. Litvinov, *Magnetism in Topological Insulators*,
https://doi.org/10.1007/978-3-030-12053-5_4

$$\nabla \cdot \boldsymbol{B} = 0$$
$$\nabla \times \boldsymbol{E} = -\frac{\partial \boldsymbol{B}}{\partial t}. \tag{4.4}$$

The second pair of Maxwell equations follows from the least action principle: conditions for minimum action by varying electromagnetic potentials which are treated as dynamical variables. The action should be invariant with respect to frame rotations in Minkowski space. The eigenvalues of F_{ik} are determined by the characteristic equation $P(\lambda) \equiv \mathrm{Det}[F_{ik} - \lambda g_{ik}] = 0$ and do not depend on the frame. This means that invariant combinations of F_{ik} can be found as coefficients in the characteristic polynomial $P(\lambda)$:

$$\mathrm{inv} = 2\left(B^2 - E^2/c^2\right) \equiv F_{ik}\, F^{ik}$$
$$\mathrm{inv} = -\frac{4\,\boldsymbol{EB}}{c} \equiv \frac{1}{2}\epsilon^{iklm} F_{ik} F_{lm}, \tag{4.5}$$

where $F^{ik} \equiv g_{il} F_{lm} g_{mk} = F_{ik}(-\boldsymbol{E}, \boldsymbol{B})$.

Taking into account both invariants (4.5), one can express the electromagnetic field action as follows:

$$S_{\mathrm{f}} = -\frac{1}{4Z_0}\int d^4x \left[F_{ik}\, F^{ik} + \frac{\kappa\,\eta}{2}\epsilon^{iklm} F_{ik} F_{lm}\right], \tag{4.6}$$

where $d^4x = d\boldsymbol{r}d(ct)$, $Z_0 = \sqrt{\mu_0/\varepsilon_0}$, μ_0, ε_0, are the impedance, permeability, and permittivity of vacuum, respectively, κ is the normalization constant to be specified later, and η is the dimensionless so-called axion field. The action of charges interacting with the electromagnetic field is written as

$$S_{\mathrm{ef}} = -\frac{1}{c}\int d^4x\, A_i j^i, \tag{4.7}$$

where the four-current is defined as $j^i = (c\rho, \boldsymbol{j})$; $\rho, \boldsymbol{j}$ are the charge and current density, respectively.

The total action from (4.6) and (4.7) can be expressed through the Lagrangian density L:

$$S = S_{\mathrm{f}} + S_{\mathrm{ef}} = \int L\, d\boldsymbol{r}dt,\;\; L = -\frac{1}{4\mu_0}\left(F_{ik}\, F^{ik} + \frac{\eta\,\kappa}{2}\epsilon^{iklm} F_{ik} F_{lm}\right) - A_i j^i. \tag{4.8}$$

Equation (4.8) is the starting point from which one can derive the second pair of Maxwell equations. Before we proceed with the derivation, let us consider the second invariant in (4.6). In three-dimensional vector form, (4.6) reads

$$S_{\mathrm{f}} = -\int drdt\left[\frac{B^2}{2\mu_0} - \frac{\varepsilon_0 E^2}{2} - \frac{\kappa\,\eta}{Z_0}\,\boldsymbol{EB}\right]. \tag{4.9}$$

If η = const, the last term in (4.9) reverses the sign under inversion $\boldsymbol{r} \to -\boldsymbol{r}$ or $t \to -t$. The term is a pseudo-scalar, i.e., it violates the Lorenz invariance and may not enter the action unless the response of the media η is a pseudo-scalar function of ($\boldsymbol{r}$, t), in which case $\eta(\boldsymbol{r}, t)$ absorbs the sign change and reinstates the symmetry of the action. This occurs in dielectrics in which both time reversal and inversion symmetry are broken. Such low-symmetry magnetic materials belong to the class that allows the linear magnetoelectric effect: the existence of the electric (magnetic) field induced magnetization (electrical polarization). In these materials the η-field is manifested as a magnetoelectric tensor that enters the thermodynamic potential [2]. The linear magnetoelectric effect in dielectrics was predicted based on symmetry considerations [2, 3], and then experimentally confirmed in Cr_2O_3 [4]. Additional data on the magnetoelectric effect in dielectrics can be found in [5, 6].

4.1 Axion Electrodynamics

The electromagnetic potential and its derivatives in (4.8) can be considered as dynamical variables which should deliver the least action. In what follows, we assume $\eta \equiv \eta(\boldsymbol{r}, t)$ and vary potentials to find conditions for the least action assuming zero variation, $\delta A^i = 0$, on the boundary of the integration region in (4.8). These conditions are the Euler–Lagrange equations:

$$\partial_i\left(\frac{\partial L}{\partial_i A_k}\right) - \frac{\partial L}{\partial A_k} = 0,\ k = 0{,}1{,}2{,}3. \tag{4.10}$$

Using L from (4.8) we present the conditions (4.10) in the form

$$-\partial_i F^{ik} - \frac{\kappa}{2}\epsilon_{iklm}\partial_i(\eta F_{lm}) + \mu_0 j^k = 0. \tag{4.11}$$

Substituting the electromagnetic tensor (4.2) into (4.11) we get the first term in (4.11) as

$$k = 0;\ \partial_i F^{ik} = \frac{1}{c}\nabla \cdot \boldsymbol{E},$$

$$k = 1;\ \partial_i F^{ik} = \partial_{x^0} F_{01} + \partial_{x^2} F_{21} + \partial_{x^3} F_{31} = -\frac{1}{c^2}\frac{\partial E_x}{\partial t} + [\nabla \times \boldsymbol{B}]_x. \tag{4.12}$$

Calculating components $k = 2, 3$ one obtains

$$\partial_i F^{ik} = \begin{cases} \frac{1}{c}\nabla \cdot \boldsymbol{E}, k = 0 \\ -\frac{1}{c^2}\frac{\partial \boldsymbol{E}}{\partial t} + \nabla \times \boldsymbol{B}, k = 1,2,3 \end{cases}. \tag{4.13}$$

The second term in (4.11):

$$\begin{aligned} k = 0;\ \epsilon_{iklm}\partial_i(\eta F_{lm}) &= \epsilon_{10lm}\partial_x(\eta F_{lm}) + \epsilon_{20lm}\partial_y(\eta F_{lm}) + \epsilon_{30lm}\partial_z(\eta F_{lm}) = \\ &= -2(\eta\,\nabla \cdot \boldsymbol{B} + \boldsymbol{B} \cdot \nabla\eta), \\ k = 1;\ \epsilon_{iklm}\partial_i(\eta F_{lm}) &= \epsilon_{01lm}\partial_{x^0}(\eta F_{lm}) + \epsilon_{21lm}\partial_y(\eta F_{lm}) + \epsilon_{31lm}\partial_z(\eta F_{lm}) = \\ &= \frac{2}{c}\left(\eta[\nabla \times \boldsymbol{E}]_x + [\nabla\eta \times \boldsymbol{E}]_x + \eta\frac{\partial B_x}{\partial t} + B_x\frac{\partial \eta}{\partial t}\right). \end{aligned} \tag{4.14}$$

Calculating terms with $k = 2, 3$ in (4.14) we gather all the components of (4.11) in vector form as

$$\begin{aligned} &\frac{\rho}{\varepsilon_0} - \nabla \cdot \boldsymbol{E} + c\,\kappa\,(\eta\nabla \cdot \boldsymbol{B} + \boldsymbol{B} \cdot \nabla\eta) = 0, \\ &\frac{1}{c^2}\frac{\partial \boldsymbol{E}}{\partial t} + \mu_0\boldsymbol{j} = \nabla \times \boldsymbol{B} - \frac{\kappa}{c}\left\{\frac{\partial(\boldsymbol{B}\eta)}{\partial t} + \eta\nabla \times \boldsymbol{E} + \nabla\eta \times \boldsymbol{E}\right\}. \end{aligned} \tag{4.15}$$

The first line in (4.15) contains the term $\eta\nabla \cdot \boldsymbol{B}$ which equals zero by virtue of (4.4). Finally, (4.4) and (4.15) present the set of Maxwell's equations modified by the axion field:

$$\begin{cases} \nabla \cdot \boldsymbol{B} = 0, \\ \nabla \times \boldsymbol{E} = -\frac{\partial \boldsymbol{B}}{\partial t}. \end{cases} \tag{4.16}$$

$$\begin{cases} \nabla \cdot \boldsymbol{E} = \frac{\rho}{\varepsilon_0} + c\,\kappa\,\boldsymbol{B} \cdot \nabla\eta, \\ \frac{1}{c^2}\frac{\partial \boldsymbol{E}}{\partial t} + \mu_0\boldsymbol{j} + \frac{\kappa}{c}\left\{\boldsymbol{B}\frac{\partial \eta}{\partial t} + \nabla\eta \times \boldsymbol{E}\right\} = \nabla \times \boldsymbol{B}. \end{cases} \tag{4.17}$$

Within classical theory, the normalization coefficient is $\kappa = 1$, while in quantum theory the interaction between electrons and the gauge field calls for $\kappa = e^2 Z_0/2\pi h = \alpha/4\pi^2\varepsilon_0$, $\alpha = e^2/\hbar c$ [7].

Once the axion field is independent of $(\boldsymbol{r}, t)$, it disappears from (4.17), making them a standard set of Maxwell equations in classical electrodynamics. When modified by the axion field η $(\boldsymbol{r}, t)$, the equations describe vacuum axion electrodynamics, where the axion field itself and physical consequences it may cause have not yet been observed experimentally.

In solid state physics the axion field is attributed to the 3D topological insulator (TI) media [8, 9] and the theory has direct physical interpretation in observable topological magnetoelectric effects.

The constant axion field violates time reversal symmetry and seems not to be compatible with the TR-symmetric bulk TI discussed in Chap. 1. In order to concentrate on compatibility we note that under time reversal the term $\boldsymbol{EB}$ in (4.9) changes the sign, thus it requires $\eta \to -\eta$ to reinstate the TR-symmetry. The operation $\eta \to -\eta$ would not change the partition function $\exp(iS/\hbar)$ (invariant under TR) if $\eta = \pi$ or $0\,(\mathrm{mod}\,2\pi)$. The quantization $(\mathrm{mod}\,2\pi)$ stems from the periodic boundary conditions [8]. So, the axion field in the bulk of the TI media is $\eta = \pi$ and zero outside it. The axion effects in (4.17) occur near the surface or domain walls where the spatial gradient $\nabla\eta$ exists. Below, we discuss the axion effects in topological insulators using phenomenological considerations based on the modified Maxwell equations (4.17).

4.2 Axion Field and Topological Polarization

To demonstrate the topological magnetoelectric effects (TME) we calculate the additional electrical charge density in (4.17) induced by the axion field on a dielectric (gapped) TI surface:

$$q_{\mathrm{a}} = \frac{e^2}{2\pi h}\boldsymbol{B}\cdot\nabla\eta. \tag{4.18}$$

First, we consider the spherical geometry illustrated in Fig. 4.1a: a magnetic monopole at $\boldsymbol{r} = 0$ (the region where $\eta = 0$), and the TI media with $\eta \neq 0$ outside the sphere of radius R. The monopole induces an electrical charge on the TI surface where the axion field gradient is $\partial\eta/\partial r = -\theta\delta(r - R)$:

$$q_{\mathrm{a}} = \frac{e^2}{2\pi h}4\pi\int_0^\infty Br^2dr\,\frac{\partial\eta}{\partial r} = -\frac{e^2\Phi\theta}{2\pi h}. \tag{4.19}$$

θ is the radial jump of the axion field on the surface and Φ is the magnetic flux. Each monopole flux quantum $\Phi = h/e$ induces a fractional electric charge on the surface of the TI:

$$q_{\mathrm{a}} = -\frac{e\theta}{2\pi} = -\frac{e}{2}. \tag{4.20}$$

The parameter θ is defined mod 2π, so the fractional charge is $q_{\mathrm{a}} = -e(n + 1/2)$, $n \in Z$. The fractional charge on a TI surface bound to the magnetic monopole is the solid state implementation of the Witten effect [9, 10].

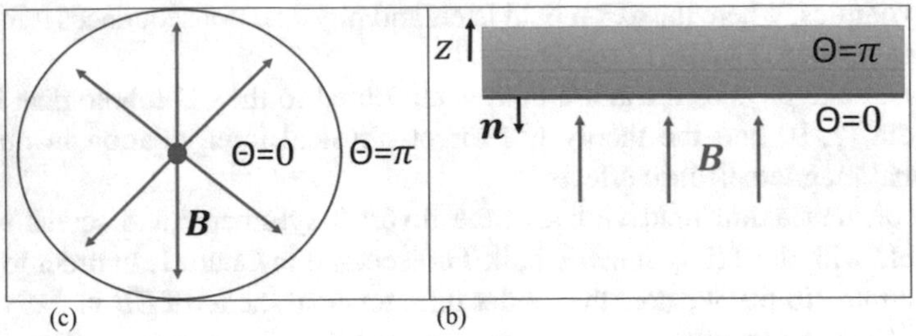

Fig. 4.1 Axion media surfaces on which the magnetic field induces electrical charges. (**a**) Magnetic monopole, (**b**) homogeneous magnetic field

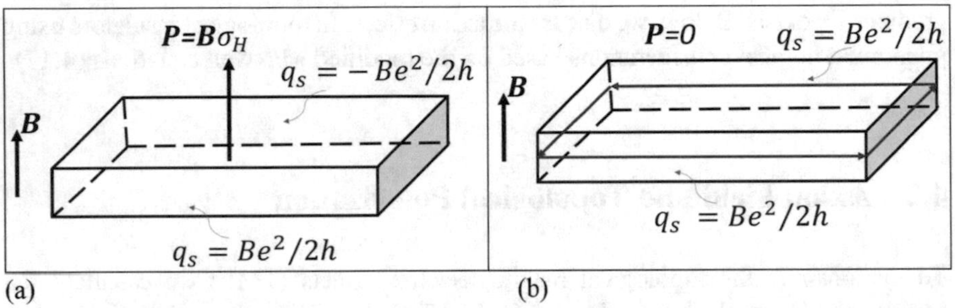

Fig. 4.2 (**a**) TME state, $\sigma_{\mathrm{H\ bottom}} = -\sigma_{\mathrm{H\ top}}$. (**b**) QHE state, $\sigma_{\mathrm{H\ bottom}} = \sigma_{\mathrm{H\ top}}$. Chiral edge current is shown in red

In plane geometry, shown in Fig. 4.1b, $\nabla\eta$ is perpendicular to the surface of the axion media, $\nabla\eta = -\boldsymbol{n}\theta\delta(z)$, $\boldsymbol{n}$ is the unit vector normal to the surface.

The magnetic field induced sheet charge density on a plane dielectric TI surface, shown in Fig. 4.1b, follows from the first line of (4.17):

$$q_{\mathrm{S}} = \frac{e^2\boldsymbol{B}}{2\pi h}\int_{-\infty}^{\infty}\nabla\eta\, dz = -\frac{e^2\boldsymbol{B}n\theta}{2\pi h} = B\sigma_{\mathrm{H}}, \sigma_{\mathrm{H}} = -\frac{e^2}{h}\left(n + \frac{1}{2}\right), \tag{4.21}$$

where σ_{H} is the quantized Hall conductivity on a single gapped TI surface.

The actual sign of conductivity cannot be specified within the phenomenological approach. Considerations based on microscopic theory and experimental settings will be discussed later in this chapter. At this point, general conclusions can be made. In a thick enough film, two independent gapped surfaces, carrying σ_{H} of opposite signs, form a double electrical layer (capacitor) with polarization equal to the sheet charge density, $\boldsymbol{P} = |\sigma_{\mathrm{H}}|\boldsymbol{B}$, manifesting the TME. The TME implies that the observable Hall conductivity, which is the sum over both surfaces, equals zero. On the other hand, the equal signs of σ_{H} on opposite surfaces result in zero polarization and a finite net QHE with $\sigma_{\mathrm{H\ total}} = \sigma_{\mathrm{top}} + \sigma_{\mathrm{bottom}}$. In other words, the QHE and TME are mutually exclusive, as illustrated in Fig. 4.2.

4.3 Axion Field and Quantized Magnetization

If one drops the time dependence ($\partial\eta/\partial t = 0$), the axion current in the second line of (4.17) is non-dissipative and is pinned to the TI surface where $\nabla\eta \neq 0$. The only way to give meaning to that kind of current is to identify it with a QHE or QAHE Hall current that flows on a gapped TI surface along the gapless outer boundary of the sample. The axion current (4.17) reads

$$\boldsymbol{j}_{\mathrm{a}} = -\frac{e^2}{2\pi h}\nabla\eta \times \mathbf{E}. \tag{4.22}$$

The direction of the axion field gradient in (4.22) is always opposite to the surface normal. Integrating (4.22) along the normal to the surface (see Fig. 4.1b) we have the sheet current density [*A*/*m*] on the surface

$$\boldsymbol{j}_{\mathrm{H}} = \frac{e^2\theta}{2\pi h}\boldsymbol{n} \times \boldsymbol{E} = \sigma_{\mathrm{H}}\boldsymbol{n} \times \boldsymbol{E}, \;\; \sigma_{\mathrm{H}} = \frac{e^2}{h}\left(n + \frac{1}{2}\right). \tag{4.23}$$

The sign of the Hall conductivity in (4.23) depends on the setting. In the QAHE, the sign is determined by the sign of magnetic gap, as follows from (3.10): $\sigma_{\mathrm{H}} = sgn(m_z)\frac{e^2}{2h}$. As $\boldsymbol{n}_{\mathrm{top}} = -\boldsymbol{n}_{\mathrm{bot}} = z$, the top and bottom Hall currents in the laboratory frame can be written as

$$\boldsymbol{j}_{\mathrm{H}} = \mathrm{sgn}(m_z)\frac{e^2}{2h}\boldsymbol{n} \times \boldsymbol{E} = \begin{cases} \sigma_{\mathrm{H\ top}}\, z \times \boldsymbol{E} \\ \sigma_{\mathrm{H\ bottom}}\, z \times \boldsymbol{E} \end{cases}. \tag{4.24}$$

In the QAHE setting, magnetic gaps of opposite signs generate the $\sigma_{\mathrm{H\ top/bottom}}$ of the same sign, and the total Hall current is determined by $\sigma_{\mathrm{H\ top}} + \sigma_{\mathrm{H\ bottom}}$. The middle of the slab (bounded by a red line in Fig. 4.2b) is the domain wall where $m_z = 0$, so gapless edge excitations propagate along the line manifesting the QAHE. This setting implies the TME is absent.

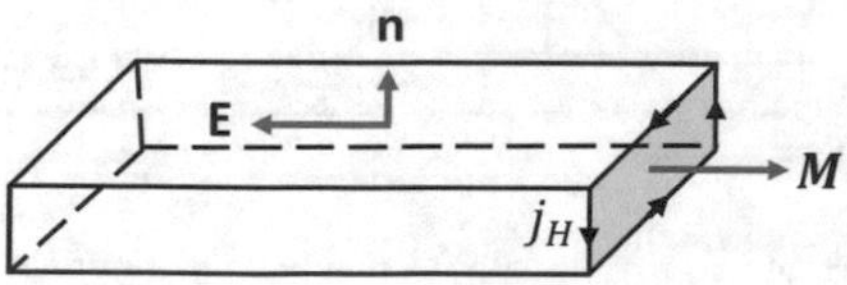

Fig. 4.3 Topological magnetic moment induced by circulating Hall current (black arrows)

4.4 TME Experimental Settings

Gapped surfaces, the Fermi level in the gap, and the total contributions from both surfaces $\sigma_{\mathrm{H\ total}} = 0$, are necessary conditions for TME to be observed. Providing all surfaces are gapped with m_z of the same sign (m_z points inward or outward over the whole 3D surface being antiparallel on the top and bottom surfaces), (4.24) guarantees opposite signs of Hall conductivities as required for the TME media. The opposite signs of j_{H} on the top and bottom surfaces result in a zero net Hall current and formation of circulating current along the sample border, as shown in Fig. 4.3.

The current shown in Fig. 4.3 circulates along the surface of a 3D TI media creating the magnetic moment $\boldsymbol{M} \sim -|\sigma_{\mathrm{H}}|\boldsymbol{E}$, another manifestation of the TME.

The TI with magnetic gaps of the same sign is illustrated in Figs. 4.2a and 4.3. The experimental implementation is achieved by covering the whole 3D TI with the same ferromagnetic material [11] or by doping top and bottom near-surface regions with magnetic atoms of different types [12], as illustrated in Fig. 4.4. Both methods result in antiparallel magnetizations on the top and bottom surfaces. We assume that the magnetic gap is positive if m_z is parallel to the surface normal, so the setting in Fig. 4.4 shows a magnetic gap of the same sign on all surfaces.

In the QHE setup, electrons on both surfaces are gapped when the Fermi level is between the successive Landau levels. The magnetic field favors parallel magnetization directions on top and bottom surfaces, so the system is in the QHE state

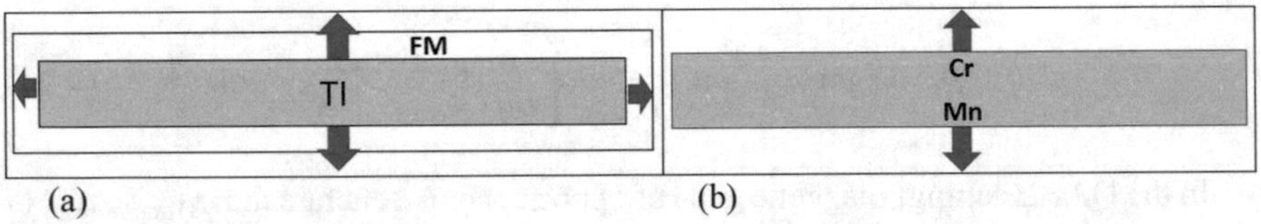

Fig. 4.4 TME setting. Same sign magnetic gap on TI surfaces. (**a**) FM is a ferromagnetic material with out-of-plane magnetic anisotropy. (**b**) Magnetic dopants with opposite signs of exchange coupling near top and bottom surfaces

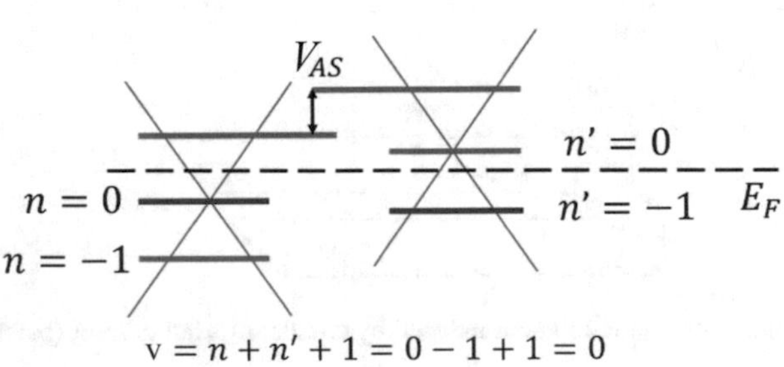

Fig. 4.5 Vertical voltage assisted $\nu = 0$ quantum Hall state. The highest occupied Landau levels (shown in red) determine the filling factor $\nu = 0$

with $\sigma_{\mathrm{H\ total}} = \sigma_{\mathrm{top}} + \sigma_{\mathrm{bottom}} = \mathrm{v}e^2/h$. Coefficient 2 in the filling factor $\mathrm{v} = 2\left(n + \frac{1}{2}\right)$ reflects the degeneracy of the top and bottom surfaces. As $\sigma_{\mathrm{H\ total}} \neq 0$, the TME is absent. The state $\mathrm{v} = 0$ would manifest the axion media where the TME is possible. It has been experimentally demonstrated that the state $\mathrm{v} = 0$ can be prepared if the top and bottom surfaces are electrically biased with respect to each other [13, 14]. As discussed in Chaps. 1 and 3, the vertical voltage $\tilde{V}_{AS}/e$ lifts the top-bottom degeneracy of the Dirac spectrum, so the filling factor, comprising top and bottom contributions, reads $\mathrm{v} = (n + n' + 1)$. The Landau level diagram in a biased film is shown in Fig. 4.5. The proper location of the Fermi level between Landau levels, shown in Fig. 4.5, brings the system into a "pseudo quantum spin Hall state" with $\mathrm{v} = 0$.

The QHE setup shown in Fig. 4.5 has been proposed for experimental verification of the TME [12]. Experimental evidence of the axion media in TI has been reported in [15] where the zero Hall state was observed in a QAHE setting.

References

1. Landau, L.D., Lifshits, E.M.: The Classical Theory of Fields, vol. 2, 4th edn. Butterworth-Heinemann (1975)
2. Landau, L.D., Lifshitz, E.M., Pitaevskii, L.P.: Electrodynamics of Continuous Media, vol. 8, 2nd edn. Butterworth-Heinemann (1984)
3. Dzyaloshinskii, I.E.: Sov. Phys. JETP. **37**, 881 (1959)
4. Astrov, D.N.: Sov. Phys. JETP (Letters). **11**, 708 (1960)
5. Fiebig, M.: J. Phys. **D38**, R123 (2005)
6. Wang, J., Neaton, J.B., Zheng, H., Nagarajan, V., Ogale, S.B., Liu, B., Viehland, D., Vaithyanathan, V., Schlom, D.G., Waghmare, U.V., Spaldin, N.A., Rabe, K.M., Wuttig, M., Ramesh, R.: Science. **299**, 1719 (2003)
7. Wilczek, F.: Phys. Rev. Lett. **58**, 1799 (1987)
8. Qi, X.-L., Hughes, T.L., Zhang, S.-C.: Phys. Rev. **195454**, B78 (2008)
9. Rosenberg, G., Franz, M.: Phys. Rev. **B82**, 035105 (2010)
10. Witten, E.: Phys. Lett. **B86**, 283 (1979)
11. Wang, J., Lian, B., Qi, X.-L., Zhang, S.-C.: Phys. Rev. **B92**, 081107(R) (2015)
12. Morimoto, T., Furusaki, A., Nagaosa, N.: Phys. Rev. **B92**, 085113 (2015)
13. Yoshimi, R., Tsukazaki, A., Kozuka, Y., Falson, J., Takahashi, K.S., Checkelsky, J.G., Nagaosa, N., Kawasaki, M., Tokura, Y.: Nat. Commun. **6**, 6627 (2015)
14. Yoshimi, R., Yasuda, K., Tsukazaki, A., Takahashi, K.S., Nagaosa, N., Kawasaki, M., Tokura, Y.: Nat. Commun. **6**, 8530 (2015)
15. Mogi, M., Kawamura, M., Yoshimi, R., Tsukazaki, A., Kozuka, Y., Shirakawa, N., Takahashi, K.S., Kawasaki, M., Tokura, Y.: Nat. Mater. **16**, 516 (2017)

Chapter 5
Rashba Effect in Topological Quantum Wells

Zero magnetic spin-splitting in quantum wells is an important tool in semiconductor spintronics. The Rashba coefficient which quantitively describes the splitting has been calculated based on a microscopic theory in two instances, both regarding topologically trivial materials: cubic [1–3] and wurtzite [4, 5] III-V semiconductors.

The near-surface region in the topological insulator Bi_2Se_3 is characterized by a large Rashba coefficient of $(0.79 - 1.3)10^{-10}$eVm [6, 7]. Topological invariants establish the connection of surface electrons to the topology of the TI bulk band structure, predicting stable and non-trivial surface states. A question arises whether and how the non-trivial bulk topology affects the bulk electronic properties of a solid. In this chapter, we consider the zero magnetic field spin-splitting of bulk origin that acquires specific features when the bulk is in the non-trivial topological phase.

5.1 Rashba Interaction in Normal Semiconductors

Spin–orbit interaction creates zero magnetic field spin-splitting in crystals and heterostructures if their symmetry does not include spatial inversion (for review, see [8, 9] and [10] for III-V cubic and wurtzite semiconductors, respectively). The theory of linear-in-k spin-splitting of electron states was developed first for wurtzite bulk materials [11–13] and then for non-centrosymmetric heterostructures [14–17]. Spin-splitting follows from symmetry considerations and can be described by the phenomenological constant called the Rashba coefficient.

The model Hamiltonian was first proposed to describe spin–orbit interaction in wurtzite semiconductors, where linear-k spin-splitting is caused by bulk inversion asymmetry:

V. Litvinov, *Magnetism in Topological Insulators*,
https://doi.org/10.1007/978-3-030-12053-5_5

$$H_R = E_c + \frac{\hbar^2 k^2}{2m} + \alpha_R \left(k_y \sigma_x - k_x \sigma_y\right) = E_c + \frac{\hbar^2 k^2}{2m} + \alpha_R [\boldsymbol{k} \times \boldsymbol{\sigma}]_z. \tag{5.1}$$

Since (5.1) accounts for the conduction band only and neglects coupling to other bands, one can consider α_R to be a phenomenological constant. Its microscopic origin is discussed in [10].

The effective mass m is introduced to account for k^2-order contributions from the remote energy bands. It should be noted that spin-orbit linear-k terms also exist in zinc-blende materials for wave vectors in the [111] and [110] directions. There, the terms appear in the expanded set of initial basis functions that accounts for d-states [18]. The Hamiltonian (5.1) is written in the spinor basis that is the pair of eigenvectors of Pauli matrix $\sigma_z : |\uparrow\rangle = \begin{pmatrix} 1 \\ 0 \end{pmatrix}, |\downarrow\rangle = \begin{pmatrix} 0 \\ 1 \end{pmatrix}$. These spinors are not eigenstates of the Hamiltonian, so the Hamiltonian is non-diagonal in spin indexes. Eigenvalues of H_R represent the spin-dependent electron energy spectrum referenced to the edge of the conduction band:

$$E = E_0 + \frac{\hbar^2 k^2}{2m} \pm \alpha_R k, \quad k = \sqrt{k_x^2 + k_y^2}, E_0 = \frac{m\alpha^2}{\hbar^2}. \tag{5.2}$$

Two branches of the spectrum are illustrated in Fig. 5.1.

Time reversal symmetry requires twofold Kramers degeneracy to hold in the spin-split one-electron spectrum: $E \uparrow (\boldsymbol{k}) = E \downarrow (-\boldsymbol{k})$. Spin-up and spin-down branches intersect at the Dirac point, $E = E_0$. The isoenergy surfaces are spherical at $E > E_0$ and toroidal at $E < E_0$. Symmetry considerations require the spin-orbit part of the total bulk wurtzite Hamiltonian to be in the form [19, 20]:

$$H_{SO} = \lambda [\boldsymbol{\sigma} \times \boldsymbol{k}]_z + \lambda_l [\boldsymbol{\sigma} \times \boldsymbol{k}]_z k_z^2 + \lambda_t [\boldsymbol{\sigma} \times \boldsymbol{k}]_z k^2, \tag{5.3}$$

where parameters λ, λ_l, λ_t are the phenomenological constants to be determined by fitting theory to experimental data. In GaN, they were found to be $\lambda < 4^*10^{-13}$eV*cm [21], and $\lambda_l \approx 4\ \lambda_t$, $\lambda_t = 7.4^*10^{-28}$eV*cm^3 [22]. The effective Hamiltonian (5.3) contains a k-linear bulk inversion asymmetry Rashba term and also high-order contributions called, by analogy to zinc-blende, k^3-Dresselhaus terms. The spin-splitting in a bulk wurtzite crystal that follows from (5.3) is absent for electrons moving in the z-direction.

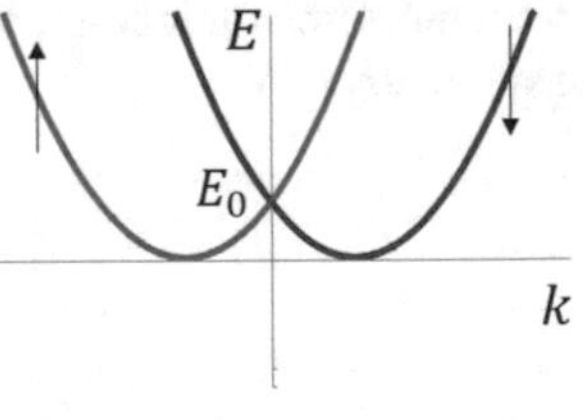

Fig. 5.1 Rashba spin-splitting. Up and down arrows stand for the two spin projections

In GaAs-based zinc-blende crystals, k-linear terms are absent, so the spin-splitting starts from the Dresselhaus k^3-terms [23]:

$$\begin{aligned} H_{\mathrm{D}} &= \delta\, \boldsymbol{\sigma} \cdot \boldsymbol{\kappa}, \\ \kappa_x &= k_x\left(k_y^2 - k_z^2\right) \\ \kappa_y &= k_y\left(k_z^2 - k_x^2\right) \\ \kappa_z &= k_z\left(k_x^2 - k_y^2\right). \end{aligned} \tag{5.4}$$

The Dresselhaus interaction (5.4) equals zero in the directions of all three principal cubic axes.

Bulk inversion asymmetry results in k-dependent zero magnetic field spin-splitting in bulk materials. No spin-splitting occurs in inversion symmetric materials like Si-Ge. Besides, the bulk Rashba and Dresselhaus coefficients in wurtzite and cubic III-V semiconductors are small and cannot be manipulated by an external force. The situation, however, looks different in low-dimensional structures like heterostructures, quantum wells, and quantum dots, as the structure inversion asymmetry may generate engineered electron spin-splitting that depends on the geometrical parameters of the structure. For instance, in quantum wells (QW), the asymmetric confining field creates structure inversion asymmetry that can be manipulated by an applied voltage as well as by the geometry of the structure. Two-dimensional electrons in asymmetric QW and (or) accumulation layers reveal linear spin-splitting and can be described by the Rashba Hamiltonian that formally coincides with that in the bulk, (5.1), where the z-axis is perpendicular to the QW plane. In an engineered asymmetric potential, the Rashba coefficient is the figure of merit for the ease with which electron spins can be modulated via an electric field. Below we compare the spin configurations induced by spin–orbit interaction in zinc-blende and wurtzite quantum wells.

Zinc-blende QW The k-linear spin–orbit interaction in zinc-blende quantum wells (ZB-QW) grown in the [001] direction comprises two contributions: the Rashba term (5.1) and the Dresselhaus term (5.4) modified by one-dimensional electron confinement:

$$H_{\mathrm{ZB}} = \alpha_{\mathrm{R}}\left(k_y\sigma_x - k_x\sigma_y\right) + \alpha_{\mathrm{D}}\left(k_x\sigma_x - k_y\sigma_y\right), \alpha_{\mathrm{D}} = -\delta\langle k_z^2\rangle, \tag{5.5}$$

where $\langle\ldots\rangle = \int(\ldots)\Phi^2 dz$, Φ is the ground state wave function of a confined electron, α_{D} and α_{R} are the Dresselhaus and Rashba coefficients, respectively. It is convenient to represent H_{ZB} in the form

$$H_{\mathrm{ZB}} = \frac{1}{\sqrt{2}}\begin{pmatrix} 0 & \gamma k \exp(i\theta) \\ \gamma k \exp(-i\theta) & 0 \end{pmatrix}, \tag{5.6}$$

where

$$\gamma = \sqrt{\alpha_{\mathrm{R}}^2 + \alpha_{\mathrm{D}}^2 + 2\alpha_{\mathrm{R}}\alpha_{\mathrm{D}} \sin 2\varphi},$$
$$\tan\theta = \frac{\alpha_{\mathrm{R}}\cos\varphi + \alpha_{\mathrm{D}}\sin\varphi}{\alpha_{\mathrm{R}}\sin\varphi + \alpha_{\mathrm{D}}\cos\varphi}, \quad \tan\varphi = k_y/k_x. \tag{5.7}$$

The diagonalized Hamiltonian and eigenvectors have the form:

$$\tilde{H} = \begin{pmatrix} -\gamma k & 0 \\ 0 & \gamma k \end{pmatrix}$$
$$u_{\pm} = \frac{1}{\sqrt{2}}\begin{pmatrix} \mp\exp(i\theta) \\ 1 \end{pmatrix}. \tag{5.8}$$

The spin-splitting following from (5.8) is $\Delta\varepsilon = 2\gamma k$. The expectation values of the spin vectors $\boldsymbol{S}_{\pm}$ in $u_{\pm}$ states can be found by calculating the matrix elements $\boldsymbol{S}_{\pm} = -\boldsymbol{S}_{\mp} = \frac{1}{2}\langle u_{\pm}|\boldsymbol{\sigma}|u_{\pm}\rangle$:

$$S_{x\pm} = \mp\frac{1}{2}\cos\theta = \mp\frac{\alpha_{\mathrm{R}}\sin\varphi + \alpha_{\mathrm{D}}\cos\varphi}{2\sqrt{\alpha_{\mathrm{R}}^2 + \alpha_{\mathrm{D}}^2 + 2\alpha_{\mathrm{R}}\alpha_{\mathrm{D}}\sin 2\varphi}},$$
$$S_{y\pm} = \pm\frac{1}{2}\sin\theta = \pm\frac{\alpha_{\mathrm{R}}\cos\varphi + \alpha_{\mathrm{D}}\sin\varphi}{2\sqrt{\alpha_{\mathrm{R}}^2 + \alpha_{\mathrm{D}}^2 + 2\alpha_{\mathrm{R}}\alpha_{\mathrm{D}}\sin 2\varphi}}. \tag{5.9}$$

The spin direction in (5.9) is linked to instantaneous electron momentum and defines the direction of the effective magnetic field $\boldsymbol{B}_{\pm}(\boldsymbol{k})$ acting on an electron. The directions of the magnetic field are opposite for electrons with opposite wave vectors or spins $(\pm)$: $\boldsymbol{B}_{\pm}(-\boldsymbol{k}) = -\boldsymbol{B}_{\pm}(\boldsymbol{k})$, $\boldsymbol{B}_{\pm}(\boldsymbol{k}) = -\boldsymbol{B}_{\mp}(\boldsymbol{k})$. The orientation of the electron spin, depending on its momentum, is illustrated in Figs. 5.2 and 5.3.

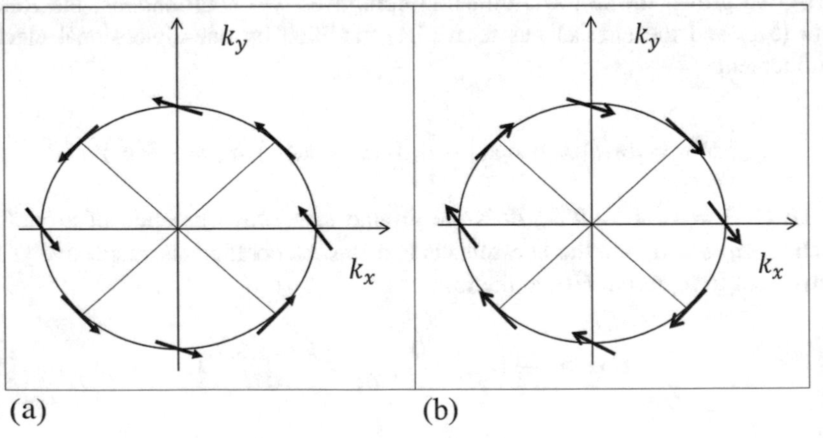

Fig. 5.2 Electron spin configuration in spin states (+) (**a**) and (−) (**b**), $\alpha_{\mathrm{D}} \neq \alpha_{\mathrm{R}}$

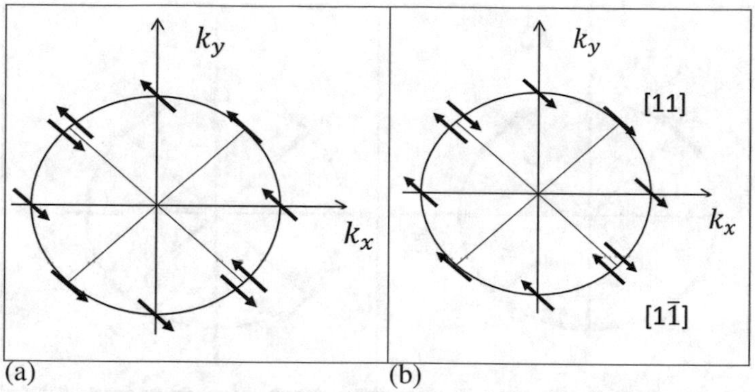

Fig. 5.3 Electron spin configuration in spin states (+) (**a**) and (−) (**b**), $\alpha_D = \alpha_R$

The conditions $\alpha_D = \pm\,\alpha_R$ make spin orientation independent of momentum in a wide area of phase space. This follows from (5.7) and is illustrated in Fig. 5.3 by the parallel spin vectors independent of the phase φ. So, the change of momentum does not rotate the spin or, in other words, the spin–orbit interaction does not preclude spin conservation against all forms of spin-independent scattering [24].

The temporal evolution of the spin texture in the case $\alpha_D = \pm\,\alpha_R$ can be reconstructed by inspecting the details in Fig. 5.3. Since the effective magnetic field is perpendicular to the [11] momentum in the (x, y) plane, the z-component of the electron spin rotates in the plane normal to (x, y) while propagating along the [11] direction. The [1$\bar{1}$] momentum direction is special as the effective magnetic field changes sign and equals zero for electrons moving in this direction. So, no spin precession occurs in the [1$\bar{1}$] direction and the z-component of electronic spin is conserved. This special case is called "persistent spin helix" [25].

When $\alpha_R \to 0$, the Dresselhaus spin texture can be found by substituting $\cos\theta \to \cos\varphi$, $\sin\theta \to \sin\varphi$. In this limit, the eigenspinors and the transformed Hamiltonian follow:

$$u_\pm = \frac{1}{\sqrt{2}}\begin{pmatrix} \mp\exp(i\varphi) \\ 1 \end{pmatrix},$$
$$\tilde{H} = \begin{pmatrix} -\alpha_D k & 0 \\ 0 & \alpha_D k \end{pmatrix}. \tag{5.10}$$

In the $u_\pm$ states, the average spin components are given as

$$S_{x\pm} = \mp\frac{1}{2}\cos\varphi; \quad S_{y\pm} = \pm\frac{1}{2}\sin\varphi, \tag{5.11}$$

and shown in Fig. 5.4.

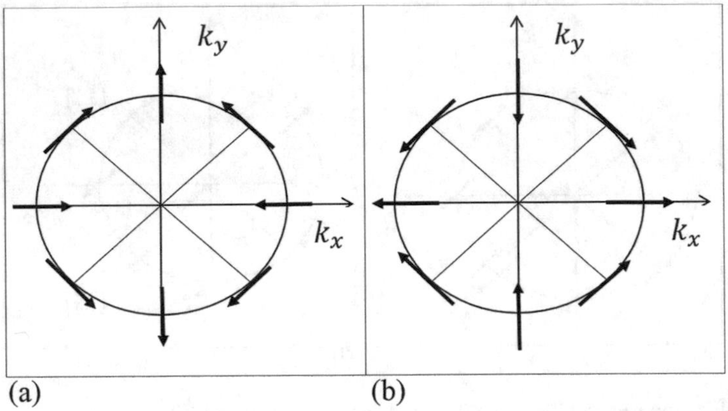

Fig. 5.4 Dresselhaus spin configurations in spin states (+) (**a**) and (−) (**b**)

Wurtzite QW In a wurtzite quantum well (W-QW) with electron confinement in the z-direction (c-axis), effective linear spin-orbit coupling follows from (5.3). Under confinement, k_z can be replaced with its quantized values k_n, $n = 1, 2...$ For instance, in the ground state the replacement can be made as

$$k_z \to \langle k_z \rangle = 0, k_z^2 \to \langle k_1^2 \rangle = (\pi/L)^2, \tag{5.12}$$

where L is the width of the infinite barrier QW. Subsequently, (5.3) can be written as

$$\begin{aligned} H_{\mathrm{W}} &= \alpha_{\mathrm{eff}}\left(\sigma_x k_y - k_x \sigma_y\right), \\ \alpha_{\mathrm{eff}} &= \alpha_{\mathrm{R}} + \alpha_{\mathrm{BIA}}, \\ \alpha_{\mathrm{BIA}} &= \lambda + \lambda_l \langle k_z^2 \rangle. \end{aligned} \tag{5.13}$$

The Hamiltonian (5.13) is a limit of (5.5) when $\alpha_{\mathrm{D}} \to 0$. The limit implies the relation between θ and the electron phase as $\cos\theta \to \sin\varphi$, $\sin\theta \to \cos\varphi$, so the eigenspinors, and transformed Hamiltonian can be written as

$$\begin{aligned} u_\pm &= \frac{1}{\sqrt{2}}\begin{pmatrix} \mp i\exp(-i\varphi) \\ 1 \end{pmatrix}, \\ \tilde{H} &= \begin{pmatrix} -\alpha_{\mathrm{eff}} k & 0 \\ 0 & \alpha_{\mathrm{eff}} k \end{pmatrix}. \end{aligned} \tag{5.14}$$

The spin-splitting is equal to $\Delta\varepsilon = 2\alpha_{\mathrm{eff}} k$. In the $u_\pm$ states, the average spin components are given as

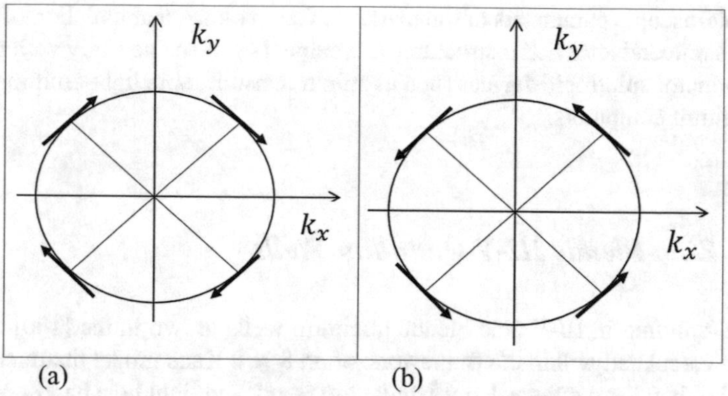

Fig. 5.5 Rashba spin configurations in spin states (+) (**a**) and (−) (**b**)

$$S_{x\pm} = \mp\frac{1}{2}\sin\varphi; \quad S_{y\pm} = \pm\frac{1}{2}\cos\varphi, \tag{5.15}$$

and their directions are shown in Fig. 5.5.

The two-dimensional (±) electron states described by the Rashba–Dresselhaus Hamiltonians are often referred in the literature as states with opposite helicity. Also, in topological insulators, electron spin states locked to the direction of propagation are referred as helical states (see Chap. 1). In Rashba and Dresselhaus electron gas, this usage is sometimes confusing, as the helicity has the exact meaning of an eigenvalue of the operator $\boldsymbol{\sigma k}/k$ that is a projection of the spin on the direction of particle momentum. It is true that sometimes the finite helicity can be associated with (±) states when spin projection on the direction of momentum is finite. However, inspecting Figs. 5.2, 5.3, 5.4, and 5.5, we can conclude that in some cases, for example, in Rashba electron gas, the helicity equals zero and cannot be a quantum number that characterizes (±) electron states. The same can be said about Rashba–Dresselhaus gas, where the helicity becomes zero in certain momentum directions. It is safe to call (±) spin states bearing in mind that they are linear combinations of the pure spin states of an electron before any spin-dependent interaction has been taken into account.

5.2 Microscopic Theory of Rashba Spin-Splitting

In heterostructures, the two-dimensional electron spectrum is spin-split if the inversion symmetry is lost due to either the asymmetric well geometry (e.g., in an accumulation layer) or to the vertical electric field originating from the external voltage applied to QW. In order to engineer spin-splitting by manipulation of QW geometrical parameters, a well/barrier lattice mismatch strain, and an external electric field, it is important to know the relation of the Rashba coefficient to the initial microscopic parameters of the material. Gate voltage manipulation of electron spins in semiconductor heterostructures is a subject of intensive study with regard to semiconductor spintronic devices such as spin transistors, spin light-emitting diodes, and quantum computers.

5.2.1 *Zinc-Blende III-V Quantum Wells*

The spin-splitting in III-V zinc-blende quantum wells grown in the [100]-direction has been calculated within the framework of an 8×8 Kane model that includes the conduction band and three valence bands: the heavy and light hole bands, as well as the spin-orbit split band [2, 3]. The Rashba coefficient for electrons is expressed as

$$\alpha = \frac{\hbar^2}{2m}\frac{\Delta}{E_g}\frac{2E_g + \Delta}{(E_g + \Delta)(3E_g + 2\Delta)}\left\langle \frac{\partial V(z)}{\partial z} \right\rangle, \tag{5.16}$$

where $V(z)/e$ is the asymmetric potential profile, m, E_g, and Δ are the effective mass, the energy gap, and Δ is the spin-orbit splitting energy, respectively. The average in (5.1) is calculated with the electron density distribution in the QW. Typical values of the Rashba coefficient α_R in InAs/GaSb, InP/InGaAs, and InAs/InGaAs QWs are $(0.6 - 4) * 10^{-11}$eVm, which gives zero magnetic field electron spin-splitting energy on the order of $(1 - 10)$ meV depending on the doping level.

5.2.2 *Rashba Coefficient in III-Nitrides*

It follows from (5.16) that the spin–orbit coupling parameter α_R decreases as the bandgap increases. Thus, spin-splitting in narrow bandgap semiconductors like InGaAs is expected to be larger than that in larger bandgap materials. However, in a wide bandgap III-Nitride heterostructure the spin-splitting, since it is approximately proportional to an average electric field in the growth direction, is affected by the in-built polarization field. Besides, the Rashba parameter α_R also depends on the Fermi level and thus a strong polarization doping effect in III-Nitrides may enhance overall spin-splitting, making it comparable to that found in zinc-blende narrow-gap III-V structures.

In heterostructures or QWs of either the AlGaN/GaN/AlGaN or GaN/InGaN/GaN-type grown in the *c*-direction, the Rashba coefficient was calculated in [4, 5]. The spin-splitting and the Rashba coefficient in a relaxed layer are given as

$$\Delta\varepsilon = 2\alpha_R k,$$
$$\alpha_R = P_1 P_2 \left\langle \frac{\partial \beta}{\partial z} \right\rangle,$$
$$\beta = \frac{\Delta_3}{(\Delta_1 - \Delta_2 + V(z) - \varepsilon)(V(z) - \varepsilon) - 2\Delta_3^2}, \tag{5.17}$$

where the reference energy is the edge of the valence band, $P_{1,\,2}$ are the momentum matrix elements, $\Delta_1$and $\Delta_{2,\,3}$ are the parameters of the crystal field and spin–orbit interactions, respectively. Here, $V(z)$ includes contributions from an external electric field (bias) and a self-consistent potential in an inhomogeneous structure.

In a generic AlGaN/GaN/AlGaN QW of width L, the coefficient β is position dependent in growth direction z and can be written as the sum of three terms each corresponding to a region i representing the left barrier l, the well w, and the right barrier r as follows:

$$\beta(z) = \beta_l[1 - \theta(z)] + \beta_w[\theta(z) - \theta(z - L)] + \beta_r\theta(z - L), \tag{5.18}$$

where $\theta(z)$ is the step function. The Rashba coefficient for the conduction band calculated with the QW ground state wave function $\Phi(z)$ is expressed as

$$\left\langle \frac{\partial \beta}{\partial z} \right\rangle = \Phi^2(0)(\beta_w - \beta_l) - \Phi^2(l)(\beta_w - \beta_r) + \langle B_l F_l \rangle + \langle B_r F_r \rangle + \langle B_w F_w \rangle, \tag{5.19}$$

where

$$\beta_i = \frac{\Delta_3}{(E_g + 2\Delta_2 + \varepsilon - V_i)(E_g + \Delta_1 + \Delta_2 + \varepsilon - V_i) - 2\Delta_3^2},$$
$$B_i = \frac{\Delta_3[2E_g + \Delta_1 + 3\Delta_2 + 2(\varepsilon - V_i)]}{\{(E_g + 2\Delta_2 + \varepsilon - V_i)(E_g + \Delta_1 + \Delta_2 + \varepsilon - V_i) - 2\Delta_3^2\}^2},$$
$$F_i = \frac{\partial V_i}{\partial z}. \tag{5.20}$$

The reference energy in (5.19) and (5.20) is the bottom of the conduction band. Each average value contains integration over the corresponding region i excluding interfaces located at $z = 0$ and $z = L$. Potential jumps and offsets of band parameters at the interfaces contribute to the first two terms of (5.19). The energy parameter ε in all B_i and β_i should be taken at the edge of the ground electron level in the well, $k = 0$, $\varepsilon \rightarrow \varepsilon_1$. All other parameters take the values attributed to the corresponding layer i.

If the barrier height tends to infinity, the coupling coefficient takes the form $\langle \partial\beta/\partial z \rangle = \langle B_w F_w \rangle$. The Rashba coefficient $\alpha_R = P_1 P_2 \langle \partial\beta/\partial z \rangle$ is proportional to the average electric field in the well if B_w does not depend on z. Otherwise, the integral in

α_R includes the product of the electric field and the combination of z-dependent material parameters. It follows from (5.19) that no spin-splitting occurs in a symmetric QW, where the following conditions hold: $\Phi^2(0)(\beta_w - \beta_l) = \Phi^2(L)(\beta_l - \beta_r)$ and $\langle B_l F_l \rangle + \langle B_r F_r \rangle = \langle B_w F_w \rangle = 0$.

Comparing the structure inversion asymmetry in zinc-blende and wurtzite materials, it is important to realize the essential difference between the two. In a symmetric and unbiased zinc-blende [100]-oriented QW, the Rashba coupling is equal to zero. On the other hand, even a structurally symmetric and unbiased wurtzite-QW is electrically asymmetric due to built-in polarization fields. This results in a nonzero Rashba coupling coefficient, which depends on internal fields. The Rashba parameter, calculated with (5.19) and (5.20), contains all of the information on the material and geometrical parameters of the structure that allow engineering the zero magnetic field spin-splitting in quasi-2D structures for various spintronic applications.

The theoretical calculations of voltage-controlled Rashba splitting in III-Nitride modulation-doped structures and quantum wells resulted in a Rashba coefficient of $\approx 2^* 10^{-13}$eVm and spin-splitting in the 1–5 meV range [4, 5]. Experimental confirmation of spin-splitting was reported in [26]. Then the spin-splitting, Rashba and Dresselhaus coefficients were measured experimentally in III-Nitride heterostructures and quantum wells of various geometry and barrier heights. The Dresselhaus coefficient (eVm^3) has been reported as $1.6^* 10^{-31}$ [27, 28] and $4^* 10^{-31}$ [29], and the Rashba coefficient (eVm) as $5.5^* 10^{-13}$ [27], $6^* 10^{-13}$ [30], $2.6^* 10^{-12}$ [24, 31, 32], $1.0^* 10^{-10}$ [33], $6.8^* 10^{-11}$ [34], $7.85^* 10^{-12}$ [35], and $4.5^* 10^{-13}$ [36]. The Rashba coefficient in bulk wurtzite GaN is less than $4^* 10^{-15}$eVm, i.e. at least two orders of magnitude smaller than those observed in quantum wells.

One more interesting feature of III-Nitrides is the narrow bandgap in InN (0.69 eV) [37] that allows an InGaN quantum well to be grown with inverted conduction and valence bands, similarly to topological insulators [38].

If polarization fields are strong enough and the Fermi energy is located as shown in Fig. 5.6, a half-metallic state exists (both electrons and holes are present in the ground state at $T = 0$). Conditions for an inverted spectrum depend on the effective bandgap in the InN well. The inverted electron spectrum is the specific feature of a topological insulator and the interplay of the Rashba interaction and the topologically non-trivial band structure is of special interest as it may result in the discovery of new spintronic capabilities of III-Nitrides. So far, in topological insulators Bi_2Se_3 the Rashba coefficient has been found to be $(0.79 - 1.3) * 10^{-10}$eVm [6, 7]. Also, a very large Rashba coefficient has been experimentally determined in BiTeJ, $(2 - 4.8) * 10^{-10}$eVm [39, 40], BiTeI [41], and BiTeI/Bi_2Te_3 heterostructures [42]. The Rashba coefficient in a system with twisted band topology is considered in the next section.

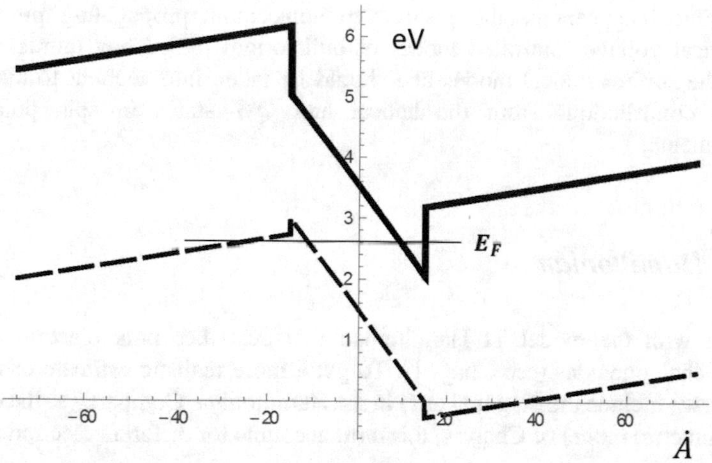

Fig. 5.6 Inverted band spectrum in a generic GaN/InN/GaN quantum well

5.3 Spin-Splitting in Topological Quantum Wells

Surface states in three-dimensional (3D) topological insulators (TI) are subject to intensive study aimed at spintronic device applications as surface electrons are gapless and at low energy can be controlled separately from the bulk insulator states. The spin-momentum locked (helical) surface states in topological insulators are at the origin of various proposals aimed at the creation and detection of surface spin polarization and spin currents [43–46]. Direct access to the helical states would make the TI a template for spintronics device applications; however, access is possible only if the chemical potential is located close to the Dirac point that is deep in the bandgap of the bulk energy spectrum. In an as-grown device, the surface defects [47] and the metal contacts [48] pin the chemical potential to the conduction or valence band and then the gate voltage is needed to control its position. The vertical bias associated with the gate voltage creates inversion asymmetry in centrosymmetric Bi_2Se_3 -based quantum wells, resulting in a spin-split QW energy spectrum. Also, the two-dimensional (2D) electron gas in an accumulation layer, which is the electrically asymmetric QW, has the spin-split spectrum which by itself creates the spin polarization of bulk origin rather than that related to the helical surface states. Discrimination between these two sources of spin polarization is important when helical states coexist with the band bending-induced QW states [49] and requires quantitative information about the spin-splitting in a TI QW.

In this section, we discuss the spin-splitting in a QW using the bulk Hamiltonian adequate to the topological state of $(Sb, Bi)_2(Te, Se)_3$, thus relating the spin-split spectrum to $k - p$ parameters of the bulk. The resulting spin-split QW spectrum determines the transport properties of TI when the Fermi level lies deep in the conduction (valence) band of the bulk and then the contribution of Dirac fermions is negligible. It appears that the spectrum contains counterpropagating spin-resolved and vertical voltage-controlled modes of bulk origin [50]. They mimic, to some extent, the surface helical modes and should be taken into account to distinguish between contributions from the helical and QW states in spin polarization measurements.

5.3.1 Hamiltonian

We start with the model TI Hamiltonian that describes bulk electron states in bismuth chalcogenides (see Chap. 1). To get a more realistic estimate of the spin-splitting, we include the diagonal term in the Hamiltonian. Compared to the electron-hole symmetric model of Chap. 1, this term accounts for different electron and hole masses:

$$H(\boldsymbol{k}) = \varepsilon(\boldsymbol{k})I + \begin{pmatrix} M(\boldsymbol{k}) - \mathrm{E} & A_1k_z & 0 & A_2k_- \\ A_1k_z & -M(\boldsymbol{k}) - \mathrm{E} & A_2k_- & 0 \\ 0 & A_2k_+ & M(\boldsymbol{k}) - \mathrm{E} & -A_1k_z \\ A_2k_+ & 0 & -A_1k_z & -M(\boldsymbol{k}) - \mathrm{E} \end{pmatrix}, \tag{5.21}$$

where the reference energy is in the middle of the bandgap, I is the 4×4 identity matrix, $\varepsilon(\boldsymbol{k}) = C + D_1k_z^2 + D_2k^2$, $M(\boldsymbol{k}) = M_0 - B_1k_z^2 - B_2k^2$, $k^2 = k_+k_-$, $k_\pm = k_x \pm ik_y$. We consider an inverted band structure, $M_0B_{1,2} > 0$. Numerical values of parameters were found from comparison to experimental data: $C = -0.0068$ eV, $D_1 = 1.3$ eVÅ^2, $D_2 = 19.6$ eVÅ^2, $M_0 = 0.28$ eV, $A_1 = 2.2$ eVÅ, $A_2 = 4.1$ eVÅ, $B_1 = 10$ eVÅ^2, $B_2 = 56$ eVÅ^2 [51]. The bulk crystal is spatial inversion symmetric and the eigenvalues of the Hamiltonian (5.21) correspond to the twofold spin degenerate conduction and valence bands: $E_{\mathrm{c,v}}(\boldsymbol{k}) = \varepsilon(\boldsymbol{k}) \pm \sqrt{M(\boldsymbol{k})^2 + A_2^2k^2 + A_1^2k_z^2}$.

Below we consider a quantum well with electron confinement in the z-direction and account for an electric field that breaks the spatial inversion symmetry in the growth direction. The field may originate either from the external vertical bias across the QW or the built-in potential in the near-surface accumulation layer in a thick film. Assuming the Fermi level lies in the conduction band, we calculate the electron spin-splitting that stems from vertically biased QW states and compare it with the experimentally observed Rashba coefficient [6].

Replacing $k_z \to -i\partial/\partial z$ in (5.21), we have the system of four equations

$$H(\boldsymbol{k})\psi = E\psi,\ \psi^{\mathrm{T}} = (u_1, u_2, u_3, u_4),$$
$$u_1 = |P1_z^+ \uparrow\rangle, u_2 = |P2_z^- \uparrow\rangle, u_3 = |P1_z^+ \downarrow\rangle, u_4 = |P2_z^- \downarrow\rangle. \tag{5.22}$$

Excluding the valence band Bloch amplitudes (u_1, u_3), one obtains the effective equation for the conduction band ($u_2 = \varphi_\uparrow$, $u_4 = \varphi_\downarrow$)

$$\tilde{H}\begin{pmatrix}\varphi_\uparrow\\ \varphi_\downarrow\end{pmatrix} = E\begin{pmatrix}\varphi_\uparrow\\ \varphi_\downarrow\end{pmatrix}, \tag{5.23}$$

where

$$\tilde{H} = \begin{pmatrix} M_0 + C + k_z\frac{\hbar^2}{m_z(E)}k_z + \frac{\hbar^2k^2}{m(E)} + V(z) & \frac{-iA_1A_2k_-}{(E+L_k)^2}\frac{\partial V(z)}{\partial z} \\ \frac{iA_1A_2k_+}{(E+L_k)^2}\frac{\partial V(z)}{\partial z} & M_0 + C + k_z\frac{\hbar^2}{m_z(E)}k_z + \frac{\hbar^2k^2}{m(E)} + V(z)\end{pmatrix},$$

$$L_k = C + M_0 + (D_2 - B_2)k^2,$$
$$\frac{\hbar^2}{m_z(E)} = \frac{A_1^2}{(E+L_k)} - B_1 + D_1,$$
$$\frac{\hbar^2}{m(E)} = \frac{A_2^2}{(E+L_k)} - B_2 + D_2. \tag{5.24}$$

The band edge potential profile $V(z) = V_{QW}(z) + V_{AS}(z)$ accounts for the symmetric QW confinement field as well as the asymmetric vertical bias (or) and the built-in potential in an accumulation layer. The expression (5.24) has been obtained using the identity

$$(-\partial_z\beta + \beta\partial_z)\varphi(z) = -\frac{\partial\beta}{\partial z}\varphi(z),$$
$$\beta = \frac{1}{E + L_k - V(z)}. \tag{5.25}$$

In (5.24), we neglect the field-dependence of the effective masses and keep the linear electric field terms in the spin-flip matrix element $\tilde{H}_{12}$. The eigenvalues of the symmetric QW follow from the diagonal part of Hamiltonian (5.24): using basis functions $\varphi_\uparrow = \varphi_\downarrow = \varphi$ one finds the twofold spin degenerate QW bands $E_{QW}(k)$ as solutions to the equation $E = \langle \tilde{H}_{11}\rangle$,

$$E = C + M_0 + \frac{\hbar^2}{m_z(E)}\langle k_z^2\rangle + \frac{\hbar^2k^2}{m(E)}. \tag{5.26}$$

The symmetric part of the potential is included in the QW solution while the asymmetric potentials is treated as a perturbation. The coefficient of the linear relation between $E_{QW}(k)$ and the electric field determines the magnitude of the spin-splitting. The spin-split spectrum in the conduction band, following from the non-diagonal terms (5.24), is given as

$$E_{\uparrow\downarrow}(k) = E_{QW}(k) \pm \alpha_R(k)k,$$
$$\alpha_R(k) = \frac{A_1 A_2}{(E_{QW}(k) + L_k)^2}\left\langle \frac{\partial V_{AS}(z)}{\partial z} \right\rangle. \tag{5.27}$$

The nonlinear equation (5.26) has two solutions for E that correspond to QW conduction and valence bands. Below we deal with the conduction band only, $E_{QW}(k)$. The numerical solutions shown below imply the L-wide infinite barrier QW, $\langle k_z^2 \rangle \approx (\pi/L)^2$, and the ground state wave function $\varphi = \sqrt{2/L}\cos(z\pi/L)$.

5.3.2 Zero Field Spin-Splitting in an Inverted Band Quantum Well

The in-plane QW dispersion in the conduction band is illustrated in Fig. 5.7. The shape of the dispersion depends on parameters. Even though the bulk conduction and valence bands are inverted in both cases (a) and (b) in Fig. 5.7, the loop of extrema in $E_{QW}(k)$ may disappear as it is sensitive to the value of spin-orbit parameter A_2 as shown in Fig. 5.7b. The normal curvature corresponds to the experimental findings reported in [49]. In what follows we keep $A_2 = 4.1$ eVÅ. If L increases, $E_{QW}(k)$ tends to the bulk branch $E_c(k, k_z = 0)$ so that the difference between the two becomes negligible at $L > 100$Å. When L decreases, the conduction and valence subbands move away from each other and transition to the direct band alignment occurs at $L < 31$Å.

The estimate of the Rashba coefficient (5.27) implies the linear asymmetric potential $V_{AS}(z) = V_g(z/L - 1/2)$, where V_g/e is the gate voltage. Spin-splitting at the gate voltage of 0.3V is shown in Fig. 5.8.

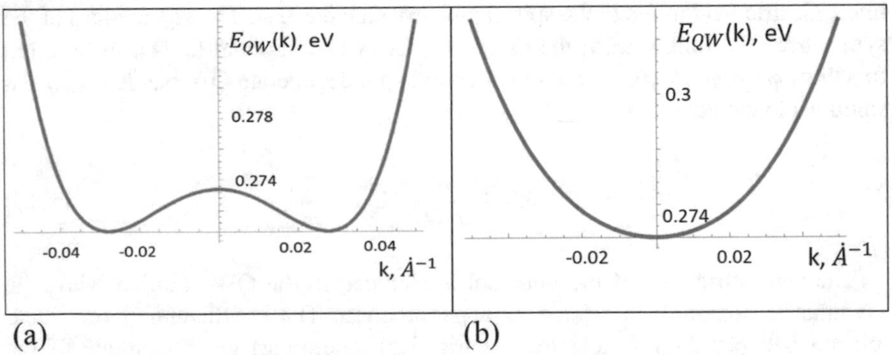

Fig. 5.7 Conduction band dispersion. QW spectrum preserves inverted band positions, $L = 45$Å. (**a**) $A_2 = 4.1$ eVÅ. (**b**) $A_2 = 5$ eVÅ

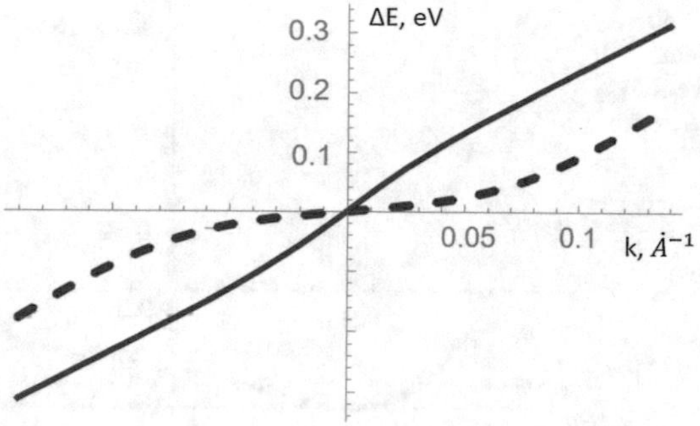

Fig. 5.8 Spin-splitting $\Delta E = E_{\uparrow}(k) - E_{\downarrow}(k)$. Solid line, $M_0 = 0.1$eV. Dashed line, $M_0 = 0.28$eV

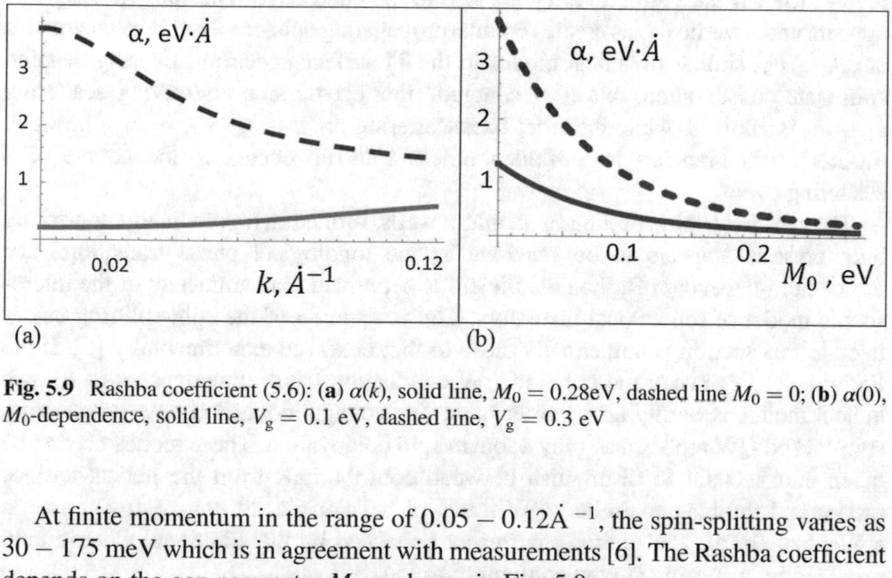

Fig. 5.9 Rashba coefficient (5.6): (**a**) $\alpha(k)$, solid line, $M_0 = 0.28$eV, dashed line $M_0 = 0$; (**b**) $\alpha(0)$, M_0-dependence, solid line, $V_g = 0.1$ eV, dashed line, $V_g = 0.3$ eV

At finite momentum in the range of $0.05 - 0.12\dot{A}^{-1}$, the spin-splitting varies as $30 - 175$ meV which is in agreement with measurements [6]. The Rashba coefficient depends on the gap parameter M_0 as shown in Fig. 5.9.

Assuming all other parameters of the Hamiltonian remain unchanged, the decrease in the gap parameter M_0 in Figs. 5.8 and 5.9 corresponds to alloying of Bi_2Se_3 with Se [52, 53] toward the topological quantum phase transition from inverted to direct band alignment. Enhanced spin-splitting in that way is illustrated in Fig. 5.8 as a comparison between two values of M_0 and in Fig. 5.9 as the explicit M_0-dependence of the Rashba coefficient. The spin-resolved QW band dispersion is illustrated in Fig. 5.10.

In the spin-resolved spectrum, presented in Fig. 5.10, the symmetry in momentum space is affected by band inversion and the momentum-dependent Rashba coefficient. The spectrum satisfies the Kramers condition $E_{\uparrow}(k) = E_{\downarrow}(-k)$ as the time reversal symmetry holds. If the Fermi level F_1 is located deep enough in the conduction band, the electron kinematics does not differ from that in the generic

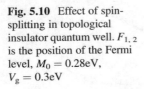

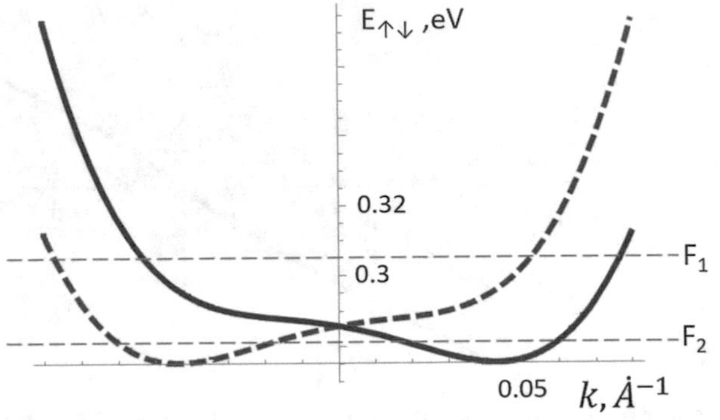

Fig. 5.10 Effect of spin-splitting in topological insulator quantum well. $F_{1,\,2}$ is the position of the Fermi level, $M_0 = 0.28\text{eV}$, $V_g = 0.3\text{eV}$

electron gas with a parabolic spectrum. Non-trivial topology of the band reveals itself if the Fermi level F_2 is close to the band edge. The spin-splitting that appears under vertical bias creates counterpropagating spin-resolved in-plane modes at $\pm k_{1,\,2}(F_2)$ similar to helical modes in the TI surface spectrum. Bearing in mind spin state conservation, one may conclude that elastic spin-conserving scattering $k_1 \to k_2$ is allowed, while the strict backscattering process $k_{1.2} \to -k_{1,\,2}$ between different spin branches is forbidden unless spin-flip occurs in the course of a scattering event.

The zero field spin-splitting in quantum wells, formed in topologically non-trivial bulk crystals, appears to be sensitive to the topological phase transition. The momentum-dependent Rashba coefficient is expressed via parameters of the microscopic model of topological insulators. The magnitude of the spin-splitting calculated in this section is numerically close to that observed experimentally [6]. If the Fermi level lies inside the inverted QW conduction band, direct access to surface helical modes is hardly achievable; however, voltage-induced counterpropagating spin-locked QW modes may play a role in spin polarization. These modes need to be taken into account to distinguish between contributions from the helical surface modes and the bias controlled QW states of bulk origin. In non-centrosymmetric solids like BiTeI, spin-splitting is further enhanced by the additional electric field provided by a specific atomic configuration [54, 55].

References

1. Pfeffer, P., Zawadski, W.: Phys. Rev. **B59**, R5312 (1999)
2. de Andrada e Silva, E.A., La Rocca, G.C., Bassani, F.: Phys. Rev. **B50**, 8523 (1994)
3. de Andrada e Silva, E.A., La Rocca, G.C., Bassani, F.: Phys. Rev. **55**, 16293 (1997)
4. Litvinov, V.I.: Phys. Rev. **B68**, 155314 (2003)
5. Litvinov, V.I.: Appl. Phys. Lett. **89**, 222108 (2006)
6. King, P.D.C., Hatch, R.C., Bianchi, M., Ovsyannikov, R., Lupulescu, C., Landolt, G., Slomski, B., Dil, H., Guan, D., Mi, J.L., Rienks, E.D.L., Fink, J., Lindblad, A., Svensson, S., Bao, S., Balakrishnan, G., Iversen, B.B., Osterwalder, J., Eberhardt, W., Baumberger, F., Hofmann, P.: Phys. Rev. Lett. **107**, 096802 (2011)

7. Zhu, Z.H., Levy, G., Ludbrook, B., Veenstra, C.N., Rosen, J.A., Comin, R., Wong, D., Dosanjh, P., Ubaldini, A., Syers, P., Butch, N.P., Paglione, J., Elfimov, I.S., Damascelli, A.: Phys. Rev. Lett. **107**, 186405 (2011)
8. Žutić, I., Fabian, J., Das Sarma, S.: Rev. Mod. Phys. **76**, 323 (2004)
9. Awshalom, D.D., Loss, D., Samarth, N.: Semiconductor Spintronics and Quantum Computation. Springer, Berlin (2002)
10. Litvinov, V.I.: Wide Bandgap Semiconductor Spintronics. Pan Stanford, Singapore (2016)
11. Rashba, E.I., Sheka, V.I.: Fiz. Tverd. Tela. **1**(2), 162 (1959)
12. Glasser, M.L.: J. Phys. Chem. Solids. **10**, 229 (1959)
13. Casella, R.C.: Phys. Rev. **114**, 1514 (1959)
14. Ohkawa, F.J., Uemura, Y.J.: J. Phys. Soc. Jpn. **37**, 1325 (1974)
15. Vasko, F.T.: Sov. Phys. JETP Lett. **30**, 541 (1979)
16. Bychkov, Y.A., Rashba, E.I.: Sov. Phys. JETP Lett. **39**, 78 (1984)
17. Bychkov, Y.A., Rashba, E.I.: J. Phys. **C17**, 6039 (1984)
18. Cardona, M., Christensen, N.E., Fasol, G.: Phys. Rev. **B38**(3), 1810–1827 (1988)
19. Bir, G.L., Pikus, G.E.: Symmetry and Strain Effects in Semiconductors. Wiley, New York (1974)
20. Zorkani, I., Kartheuser, E.: Resonant magnetooptical spin transitions in zinc-blende and wurtzite semiconductors. Phys. Rev. **B53**(4), 1871–1880 (1996)
21. Wołoś, A., Wilamowski, Z., Skierbiszewski, C., Drabinska, A., Lucznik, B., Grzegory, I., Porowski, S.: Phys. B. **406**(13), 2548–2554 (2011)
22. Wang, W.T., Wu, C.L., Tsay, S.F., Gau, M.H., Lo, I., Kao, H.F., Jang, D.J., Chiang, J.C., Lee, M.E., Chang, Y.C., Chen, C.N., Hsueh, H.C.: Appl. Phys. Lett. **91**, 082110 (2007)
23. Dresselhaus, G.: Phys. Rev. **100**, 580–586 (1955)
24. Schliemann, J., Egues, J.C., Loss, D.: Phys. Rev. Lett. **90**(14), 146801 (2003)
25. Koralek, J.D., Weber, C.P., Orenstein, J., Bernevig, B.A., Zhang, S.C., Mack, S., Awschalom, D.D.: Nature. **458**, 610–614 (2009)
26. Weber, W., Ganichev, S.D., Danilov, S.N., Weiss, D., Prettl, W.: Appl. Phys. Lett. **87**, 262106 (2005)
27. Kurdak, Ç., Biyikli, N., Özgür, Ü., Morkoç, H., Litvinov, V.I.: Phys. Rev. **B74**, 113308 (2006)
28. Cheng, H., Biyikli, N., Özgür, Ü., Kurdak, Ç., Morkoç, H., Litvinov, V.I.: Physica. **E40**, 1586–1589 (2008)
29. Yin, C., Shen, B., Zhang, Q., Xu, F., Tang, N., Cen, L., Wang, X., Chen, Y., Yu, J.: Appl. Phys. Lett. **97**, 181904 (2010)
30. Schmult, S., Manfra, M.J., Punnoose, A., Sergent, A.M., Baldwin, K.W., Molnar, R.: Phys. Rev. **B74**, 033302 (2006)
31. Cho, K.S., Liang, C.T., Chen, Y.F., Fan, J.C.: Semicond. Sci. Technol. **22**, 870–874 (2007)
32. Spirito, D., Frucci, G., Di Gaspare, A., Di Gaspare, L., Giovine, E., Notargiacomo, A., Roddaro, S., Beltram, F., Evangelisti, F.: J. Nanopart. Res. **13**, 5699–5704 (2011)
33. Belyaev, A.F., Raicheva, V.G., Kurakin, A.M., Klein, N., Vitusevich, S.A.: Phys. Rev. **B77**, 035311 (2008)
34. Zhou, W.Z., Lin, T., Shang, L.Y., Sun, L., Gao, K.H., Zhou, Y.M., Yu, G., Tang, N., Han, K., Shen, B., Guo, S.L., Gui, Y.S., Chu, J.H.: J. Appl. Phys. **104**(5), 053703 (2008)
35. Lisesivdin, S.B., Balkan, N., Makarovsky, O., Patanè, A., Yildiz, A., Caliskan, M.D., Kasap, M., Ozcelik, S., Ozbay, E.: J. Appl. Phys. **105**, 093701 (2009)
36. Stefanowicz, W., Adhikari, R., Andrearczyk, T., Faina, B., Sawicki, M., Majewski, J.A., Dietl, T., Bonanni, A.: Phys. Rev. **B89**, 205201 (2014)
37. Walukiewicz, W., Li, S.X., Wu, J., Yu, K.M., Ager III, J.W., Haller, E.E., Lu, H., Schaff, W.J.: J. Cryst. Growth. **269**, 119–127 (2004)
38. Miao, M.S., Yan, Q., Van de Walle, C.G., Lou, W.K., Li, L.L., Chang, K.: Phys. Rev. Lett. **109**, 186803 (2012)
39. Eremeev, S.V., Nechaev, I.A., Koroteev, Y.M., Echenique, P.M., Chulkov, E.V.: Phys. Rev. Lett. **108**, 246802 (2012)

40. Rusinov, P., Nechaev, I.A., Eremeev, S.V., Friedrich, C., Blügel, S., Chulkov, E.V.: Phys. Rev. **B87**, 205103 (2013)
41. Ishizaka, K., Bahramy, M.S., Murakawa, H., Sakano, M., Shimojima, T., Sonobe, T., Koizumi, K., Shin, S., Miyahara, H., Kimura, A., Miyamoto, K., Okuda, T., Namatame, H., Taniguchi, M., Arita, R., Nagaosa, N., Kobayashi, K., Murakami, Y., Kumai, R., Kaneko, Y., Onose, Y., Tokura, Y.: Nat. Mater. **10**, 521–526 (2011)
42. Zhou, J.J., Feng, W., Zhang, Y., Yang, S.A., Yao, Y.: Sci. Rep. **4**, 3841 (2014)
43. Li, C.H., van 't Erve, O.M.J., Robinson, J.T., Liu, Y., Li, L., Jonker, B.T.: Nat. Nanotechnol. **9**, 218–224 (2014)
44. Li, C.H., van't Erve, O.M.J., Rajput, S., Li, L., Jonker, B.T.: Nat. Commun. **7**, 13518 (2016)
45. Dankert, A., Geurs, J., Kamalakar, M.V., Charpentier, S., Dash, S.P.: Nano. Lett. **15**, 7976–7981 (2015)
46. Burkov, A.A., Hawthorn, D.G.: Phys. Rev. Lett. **105**, 066802 (2010)
47. Suh, J., Fu, D., Liu, X., Furdyna, J.K., Yu, K.M., Walukiewicz, W., Wu, J.: Phys. Rev. **B89**, 115307 (2014)
48. Spataru, C.D., Leonard, F.: Phys. Rev. **B90**, 085115 (2014)
49. Bianchi, M., Guan, D., Bao, S., Mi, J., Iversen, B.B., King, P.D.C., Hofmann, P.: Nat. Commun. **1**, 128 (2010)
50. Litvinov, V.I.: Phys. Rev. **B97**, 235438 (2018)
51. Zhang, H., Liu, C.-X., Qi, X.-L., Dai, X., Fang, Z., Zhang, S.-C.: Nat. Phys. **5**, 438–442 (2009)
52. Zhang, J., Chang, C.Z., Tang, P., Zhang, Z., Feng, X., Li, K., Wang, L., Chen, X., Liu, C., Duan, W., He, K., Xue, Q.K., Ma, X., Wang, Y.: Science. **339**, 1586 (2013)
53. Zhang, Q., Zhang, Z., Zhu, Z., Schwingenschlögl, U., Cui, Y.: ACS Nano. **6**, 2345 (2012)
54. Bahramy, M.S., Arita, R., Nagaosa, N.: Phys. Rev. **B84**, 041202(R) (2011)
55. Crepaldi, A., Moreschini, L., Autes, G., Tournier-Colletta, C., Moser, S., Virk, N., Berger, H., Bugnon, P., Chang, Y.J., Kern, K., Bostwick, A., Rotenberg, E., Yazyev, O.V., Grioni, M.: Phys. Rev. Lett. **109**, 096803 (2012)

Chapter 6
Spin-Electron (*s-d*) Interaction in TI Thin Films

Conducting surface states in topological insulators (TIs) present a playground for studying the response of massless Dirac fermions to various perturbations such as applied fields, ternary alloy variation, magnetic doping, as well as superconducting and ferromagnetic proximity effects. The ability to control the magnetic properties of surface electrons is key to the technology that might bring TI into the current discourse of spintronic device applications. The way surface ferromagnetism acts on TI is that it opens an energy gap in the Dirac spectrum on a single surface. Ferromagnetism itself is affected by the surface electrons as the indirect exchange interaction, mediated by Dirac electrons and holes, may or may not cause ferromagnetic ordering, depending on the details of the energy spectrum.

One of the approaches to studying the indirect exchange is based on a model that describes massless Dirac electrons interacting with localized spins by contact *s-d* interaction [1–6]. This model is simplistic in the sense that it is comprised of two terms of different origins: the Dirac model which is an effective theory obtained by projection of the bulk Hamiltonian onto surface states (see Chap. 1) and the *s-d* interaction on the surface which is normally assumed to be equivalent to that in the bulk. As a result, the *s-d* interaction constants are not specific to surface states in that they do not depend on the position of localized spin while one would expect the interaction to fade with the distance from the surface. This factor might be important as in real materials the magnetic ions are not pinned to a surface, but are rather distributed in the whole sample, making the simplified model not adequate for real situation. Besides, most models account for a single Dirac cone, and are thus not applicable to a thin film where the two Dirac cones on opposite surfaces couple with each other.

A consistent approach to surface magnetism should operate with the effective *s-d* coupling calculated as a projection of the bulk Hamiltonian that comprises electrons, localized spins, and the contact *s-d* interaction between them. In this chapter we discuss an effective surface model which serves as a background for the analytical study of surface magnetism [7]. Within the effective surface model, the interaction matrix depends on the position of localized spins relative to the slab surfaces.

V. Litvinov, *Magnetism in Topological Insulators*,
https://doi.org/10.1007/978-3-030-12053-5_6

It results in variable impurity spin textures if we move away from the surfaces. Also, *s-d* interaction constants depend on the parameters of the electron spectrum that makes the constants specific to a particular TI whether it is a thick sample with a single massless Dirac cone or a thin slab with gapped massive fermions. The interaction matrix is a function of the gate bias and is affected by the Rashba spin splitting experimentally observed in TI [8]. The effective spin-electron interaction determines the indirect exchange coupling that is at the origin of the impurity ferromagnetism to be discussed in Chap. 8.

6.1 Contact *s-d* Interaction in Metals

In metals, the indirect exchange appears as a result of the coupling between the impurity and free *s*-electron spin moments. The localized spin originates from the unfilled *d*-shell of a transition metal atom or the *f*-shell of a rare earth element, so the coupling is referred to as *s-d* or *s-f* interaction. As the inner unfilled orbitals are more localized than the valence *s*-states, the *s-d* interaction occurs in close proximity to a magnetic impurity and can be approximated by a contact interaction between the electron spin $\boldsymbol{\sigma}$ and the spin of a magnetic atom $\boldsymbol{S}$ located at point $\boldsymbol{R}_0$ [9, 10]:

$$H_{\mathrm{sd}} = \frac{J}{n} \boldsymbol{S}\boldsymbol{\sigma}\delta(\boldsymbol{r} - \boldsymbol{R}_0), \tag{6.1}$$

where n is the volume density of host atoms. In a single-band metal, the *s-d* interaction can be represented in the second quantization form by expanding the operator wave function in plane waves normalized on the crystal volume

$$\Psi(\boldsymbol{r}) = \frac{1}{\sqrt{V}} \sum_{\boldsymbol{k},\, s} a_{\boldsymbol{k}s} \exp(i\boldsymbol{k}\boldsymbol{r}) \tag{6.2}$$

and then calculating the matrix element of the interaction (6.1):

$$\tilde{H}_{\mathrm{sd}} = \int \Psi^{+}(\boldsymbol{r}) H_{\mathrm{sd}} \Psi(\boldsymbol{r}) d\boldsymbol{r} = \frac{J}{N} \sum_{\boldsymbol{k},\boldsymbol{q}} \sum_{s,\, s'} \exp(i\boldsymbol{q}\boldsymbol{R}_0) \boldsymbol{S}\boldsymbol{\sigma}_{s,s'} a^{+}_{\boldsymbol{k}s} a_{\boldsymbol{k}+\boldsymbol{q},s'}, \tag{6.3}$$

where $a_{\boldsymbol{k}s}$ is the annihilation operator of an electron with spin $s = \pm 1/2$, $\boldsymbol{\sigma}_{s,s'}$ are Pauli matrices, and N is the number of host atoms.

The electron scattering amplitude in metals, deduced from interaction (6.3), experiences logarithmic divergence at the Fermi energy if $J > 0$, and leads to an increase in electrical resistivity with decreasing temperature (Kondo effect [9]). Also, interaction (6.3), treated as a perturbation up to the second order, generates indirect exchange coupling between a pair of impurity spins. In a simple metal the coupling is the Ruderman–Kittel–Kasuya–Yosida (RKKY) interaction [11–13]. Detailed calculation of the indirect exchange interaction in metals of various dimensions and semiconductors can be found in [14, 15], respectively.

6.2 Contact *s-d* Interaction in TI Films

The basis wave functions in the semiconductor host $Bi_2(Se,Te)_3$ are more complex than the single-band plane waves used in (6.3). As discussed in Chap. 1, the set of basis wave functions in bulk Bi_2Te_3 is the 4-spinor comprising bulk Bloch amplitudes at $\boldsymbol{k} = 0 : u^T = \{(v\uparrow),(c\uparrow),(v\downarrow),(c\downarrow)\}$. Taking into account that surfaces are perpendicular to the z-direction, the operator wave function can be expanded similar to (6.2):

$$\Psi(\boldsymbol{r}) = \frac{1}{\sqrt{A}}\sum_{kj} a_{kj}u_j \exp\left(i\boldsymbol{k}\boldsymbol{r}_\parallel\right), \quad \boldsymbol{r}_\parallel = (x,y), \;\; \boldsymbol{k} = \left(k_x,k_y\right), \tag{6.4}$$

where A is the film area. The electron part of the Hamiltonian in this representation is given as

$$H = \sum_{i,j} H_{ij}(\boldsymbol{k})\, a^+_{ik} a_{jk},$$
$$H_{ij} = [V_S(z) + V_{AS}(z)]I$$
$$+\begin{pmatrix} -\Delta + B_1\frac{\partial^2}{\partial z^2} - B_2k^2 & -iA_1\frac{\partial}{\partial z} & 0 & A_2k_- \\ -iA_1\frac{\partial}{\partial z} & \Delta - B_1\frac{\partial^2}{\partial z^2} + B_2k^2 & A_2k_- & 0 \\ 0 & A_2k_+ & -\Delta + B_1\frac{\partial^2}{\partial z^2} - B_2k^2 & iA_1\frac{\partial}{\partial z} \\ A_2k_+ & 0 & iA_1\frac{\partial}{\partial z} & \Delta - B_1\frac{\partial^2}{\partial z^2} + B_2k^2 \end{pmatrix}. \tag{6.5}$$

In order to compose the full Hamiltonian including the magnetic interaction, we have to express the *s-d* coupling (6.1) in the same representation as the Hamiltonian (6.5):

$$H_{\rm sd} = \int \Psi^+(\boldsymbol{r}_\parallel) H_{\rm sd} \Psi(\boldsymbol{r}_\parallel) d\boldsymbol{r}_\parallel = \frac{1}{nA} \sum_{ki,k'j} J_{ij} \exp\left(i(\boldsymbol{k}-\boldsymbol{k}')\boldsymbol{R}_{0\parallel}\right) \delta(z-Z) a^+_{i\boldsymbol{k}}\, a_{j\boldsymbol{k}'},$$

$$J_{ij} = \begin{pmatrix} J_{\rm v} S_z & 0 & J_{\parallel{\rm v}} S_- & 0 \\ 0 & J_c S_z & 0 & J_{\parallel c} S_- \\ J_{\parallel{\rm v}} S_+ & 0 & -J_{\rm v} S_z & 0 \\ 0 & J_{\parallel c} S_+ & 0 & -J_c S_z \end{pmatrix},$$

$$J_i = J u^*_{i\uparrow}(\boldsymbol{R}_{0\parallel}) u_{i\uparrow}(\boldsymbol{R}_{0\parallel}) = J u^*_{i\downarrow}(\boldsymbol{R}_{0\parallel}) u_{i\downarrow}(\boldsymbol{R}_{0\parallel}),$$

$$J_{\parallel i} = J u^*_{i\uparrow}(\boldsymbol{R}_{0\parallel}) u_{i\downarrow}(\boldsymbol{R}_{0\parallel}), \quad i = c, {\rm v}. \tag{6.6}$$

In what follows we assume that coupling constants $J_{c,{\rm v}}$ in (6.6) are not dependent on the in-plane position of the localized spin. The z-coordinate is kept separate to account for finite size effects in the z-direction, $Z \equiv \boldsymbol{R}_{0z}$, $S_\pm = S_x \pm i S_y$. The *s-d* interaction matrix (6.6) is obtained under the assumption that inter-band matrix elements are much smaller than intra-band ones and can be neglected.

Using surface basis wave functions one could obtain an effective Hamiltonian for surface states (Chap. 1). The same procedure should be applied to the full Hamiltonian $H + H_{\rm sd}$ in order to get the full effective Hamiltonian that includes the interaction of surface electrons with an impurity spin:

$$H_{\rm eff} = \int_{-L/2}^{L/2} \Phi_i^+ (H + H_{\rm sd}) \Phi_j dz = H^0_{\rm eff} + H^1_{\rm eff} + H^{\rm sd}_{\rm eff}. \tag{6.7}$$

The first two terms in (6.7) were obtained in Chap. 1:

$$H^0_{\rm eff} = E_0 + Dk^2 + \tilde{V}_s,$$

$$H^1_{\rm eff} = \begin{pmatrix} \Delta_S/2 - Bk^2 & \tilde{V}_{as} & 0 & i\tilde{A}_2 k_- \\ \tilde{V}_{as} & -\Delta_S/2 + Bk^2 & i\tilde{A}_2 k_- & 0 \\ 0 & -i\tilde{A}_2 k_+ & \Delta_S/2 - Bk^2 & \tilde{V}_{as} \\ -i\tilde{A}_2 k_+ & 0 & \tilde{V}_{as} & -\Delta_S/2 + Bk^2 \end{pmatrix}. \tag{6.8}$$

The effective surface *s-d* interaction is

$$H^{\rm sd}_{\rm eff} = \int_{-L/2}^{L/2} \Phi_i^+ H_{\rm sd} \Phi_j dz. \tag{6.9}$$

One more simplification is possible if energy bands possess electron-hole symmetry. This reduces the number of coupling constants: $J_{\rm v} = J_c \equiv J_z$, $J_{\parallel{\rm v}} = J_{\parallel c} \equiv J_\parallel$. After that the effective Hamiltonian (6.9) can be calculated directly making use of wave functions given in (1.28):

$$H_{\text{eff}}^{\text{sd}} = \frac{1}{nA} \sum_{ki, k'j} W_{ij}(Z) \exp\left(i(\boldsymbol{k} - \boldsymbol{k}')\boldsymbol{R}_{\boldsymbol{0}\parallel}\right) a_{ik}^{+} a_{jk'}, \tag{6.10}$$

where

$$W(Z) = \begin{pmatrix} J_{z1}S_z & J_{z2}S_z & J_t S_- & 0 \\ J_{z2}S_z & J_{z1}S_z & 0 & -J_t S_- \\ J_t S_+ & 0 & -J_{z1}S_z & -J_{z2}S_z \\ 0 & -J_t S_+ & -J_{z2}S_z & -J_{z1}S_z \end{pmatrix},$$

$$\begin{gathered} J_{z1}/J_z = \varphi_\uparrow^+(Z)\varphi_\uparrow(Z) = \varphi_\downarrow^+(Z)\varphi_\downarrow(Z) = \chi_\uparrow^+(Z)\chi_\uparrow(Z) = \chi_\downarrow^+(Z)\chi_\downarrow(Z), \\ J_{z2}/J_z = \varphi_\uparrow^+(Z)\chi_\uparrow(Z) = \varphi_\downarrow^+(Z)\chi_\downarrow(Z), \\ J_t/J_\parallel = \varphi_\uparrow^+(Z)\varphi_\downarrow(Z) = \chi_\uparrow^+(Z)\chi_\downarrow(Z). \end{gathered} \tag{6.11}$$

Formally, the matrix (6.11) should include parameters proportional to $\varphi_\downarrow^+(Z)\chi_\uparrow(Z)$; however, compared to those defined in (6.11), they are negligibly small and are not considered here. The Hamiltonian (6.11) presents an effective model for surface electrons interacting with a localized spin in a TI film. The *s-d* interaction matrix is *Z*-dependent and reflects the localized nature of surface states. Matrix elements as functions of magnetic atom position in the direction normal to surfaces are illustrated in Fig. 6.1. The structure of *s-d* interaction of surface electrons is more complicated than initial bulk interaction (6.6) as it contains the inter-band transitions originating from the surface-induced mixing of initial basis functions.

The spin-flip constant J_t tends to zero in the vicinity of either of the two surfaces, while it is proportional to the overlap of the top and bottom wave functions in the middle of the film, being much smaller than diagonal *s-d* parameters. The numerical data shown in Fig. 6.1 were obtained for a slab thickness of four quintuple layers. For a thinner slab, the spin-flip constant in the middle of the slab increases, remaining ten times smaller than the diagonal non-spin-flip parameters.

In order to better understand the behavior of *s-d* parameters as functions of the *z*-position of the localized spin, it is instructive to look at the interaction matrix in the top(*t*)-bottom(*b*) representation given in (1.47). The transition to the top-bottom basis states $\{b\uparrow, b\downarrow, t\uparrow, t\downarrow\}$ is carried out by the unitary transformation

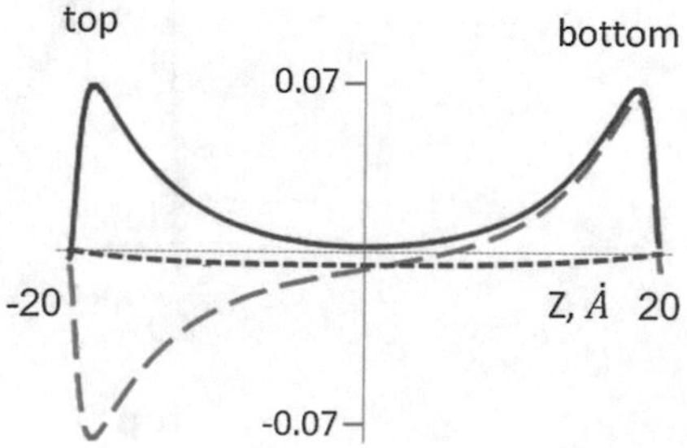

Fig. 6.1 *s-d* constants from (6.11) as functions of a localized spin position across the film thickness (*z*-direction). Functions $\varphi(z)$, $\chi(z)$ are normalized on the film thickness. Solid line—J_{z1}/J_z, dashed line—J_{z2}/J_z, dotted line—$J_t/J_\parallel$

$$U = \frac{1}{\sqrt{2}} \begin{pmatrix} 1 & 1 & 0 & 0 \\ 0 & 0 & 1 & 1 \\ -1 & 1 & 0 & 0 \\ 0 & 0 & -1 & 1 \end{pmatrix} \tag{6.12}$$

applied to the Hamiltonian (6.7). As a result, the full Hamiltonian is expressed as

$$(H_{\text{eff}})_{tb} = \left(H^0_{\text{eff}} + H^1_{\text{eff}} + H^{\text{sd}}_{\text{eff}}\right)_{tb},$$

$$\left(H^0_{\text{eff}} + H^1_{\text{eff}}\right)_{tb} = H^0_{\text{eff}} + \begin{pmatrix} \tilde{V}_{AS} & i\tilde{A}_2 k_- & -\frac{\Delta_S}{2} + Bk^2 & 0 \\ -i\tilde{A}_2 k_+ & \tilde{V}_{AS} & 0 & -\frac{\Delta_S}{2} + Bk^2 \\ -\frac{\Delta_S}{2} + Bk^2 & 0 & -\tilde{V}_{AS} & -i\tilde{A}_2 k_- \\ 0 & -\frac{\Delta_S}{2} + Bk^2 & i\tilde{A}_2 k_+ & -\tilde{V}_{AS} \end{pmatrix},$$

$$\left(H^{\text{sd}}_{\text{eff}}\right)_{tb} = \begin{pmatrix} J_{\text{bot}} S_z & 0 & 0 & -J_t S_- \\ 0 & -J_{\text{bot}} S_z & -J_t S_+ & 0 \\ 0 & -J_t S_- & J_{\text{top}} S_z & 0 \\ -J_t S_+ & 0 & 0 & -J_{\text{top}} S_z \end{pmatrix}. \tag{6.13}$$

Parameters $J_{\text{bot}} = J_{z1} + J_{z2}$ and $J_{\text{top}} = J_{z1} - J_{z2}$ in (6.13) describe non-spin-flip *s-d* interaction on the bottom and top surfaces, respectively. As shown in Fig. 6.2, they fade out when the impurity moves away from the respective surface.

Inspecting the *s-d* matrix (6.13) one comes to the conclusion that the spin-flip electron scattering against localized magnetic impurity in TI film is suppressed as it accompanies top-bottom electron transitions with much lower amplitude than the

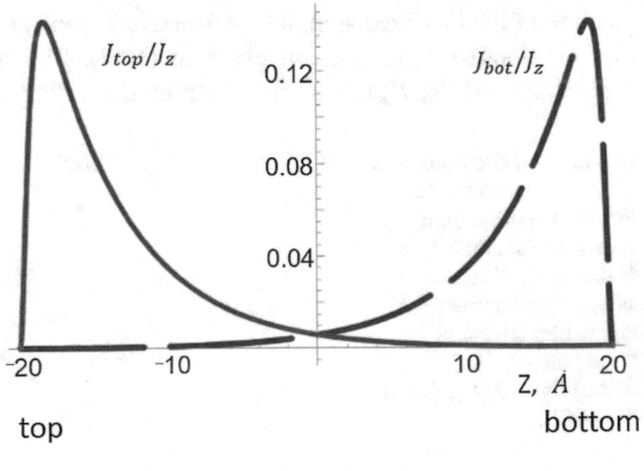

Fig. 6.2 *s-d* exchange parameters vs. *Z*-position of a magnetic impurity: dashed line, J_{bot}/J_z, solid line, J_{top}/J_z

strong non-spin-flip processes in the vicinity of each surface. In a thick enough film, the parameter J_t is negligible and the diagonal *s-d* matrix (6.13) describes the *s-d* interaction on independent surfaces.

6.3 Real Fermions Interacting with Magnetic Impurity

We deal with the Hamiltonian that consists of the electron part (6.8) and the *s-d* interaction part given in (6.10). The basis functions used make the *s-d* constants independent of electron momentum; however, as the basis does not comprise eigenstates, the electron part of the Hamiltonian is non-diagonal. Let us consider the shape of the *s-d* interaction in a representation which diagonalizes the electronic part of the Hamiltonian. The new basis composed of eigenstates of (6.8),

$$
\begin{aligned}
C_{\uparrow\downarrow} &= \sqrt{\frac{Bk^2-\Delta_S/2+R_{\uparrow\downarrow}}{4R_{\uparrow\downarrow}}}\begin{pmatrix} \dfrac{(\tilde{A}_2k \mp \tilde{V}_{AS})i\exp(-i\varphi)}{Bk^2-\Delta_S/2+R_{\uparrow\downarrow}} \\ \mp i\exp(-i\varphi) \\ \dfrac{\tilde{V}_{AS} \mp \tilde{A}_2k}{Bk^2-\Delta_S/2+R_{\uparrow\downarrow}} \\ 1 \end{pmatrix}, \\
V_{\uparrow\downarrow} &= \sqrt{\frac{\Delta_S/2-Bk^2+R_{\uparrow\downarrow}}{4R_{\uparrow\downarrow}}}\begin{pmatrix} \dfrac{(\tilde{A}_2k \mp \tilde{V}_{AS})i\exp(-i\varphi)}{\Delta_S/2-Bk^2+R_{\uparrow\downarrow}} \\ \mp i\exp(-i\varphi) \\ \dfrac{\tilde{V}_{AS} \mp \tilde{A}_2k}{\Delta_S/2-Bk^2+R_{\uparrow\downarrow}} \\ 1 \end{pmatrix}
\end{aligned}
\tag{6.14}
$$

diagonalizes the Hamiltonian (6.8), resulting in a diagonal matrix where the main diagonal comprises four branches of the spectrum of real fermions. The term "real" here means that these four branches are observed in angle-resolved photoemission spectroscopy measurements:

$$
\begin{aligned}
H_{\mathrm{el}} &= \mathrm{diag}\{E_{\mathrm{v}\uparrow}, E_{\mathrm{c}\uparrow}, E_{\mathrm{v}\downarrow}, E_{\mathrm{c}\downarrow}\}, \\
E_{\mathrm{c,v}\uparrow} &= \tilde{V}_S + Dk^2 \pm R_\uparrow, \\
E_{\mathrm{c,v}\downarrow} &= \tilde{V}_S + Dk^2 \pm R_\downarrow, \\
R_{\uparrow\downarrow} &= \sqrt{\left(\tfrac{\Delta_S}{2} - Bk^2\right)^2 + \left(\tilde{A}_2k \pm \tilde{V}_{AS}\right)^2}\,.
\end{aligned}
\tag{6.15}
$$

Presenting the *s-d* part of the Hamiltonian in the same basis one gets the *s-d* interaction between real fermions and the magnetic atom placed at point ($\boldsymbol{R}_{0\|}$, Z).

$$\tilde{H}^{\text{sd}}_{\text{eff}} = \frac{1}{nA}\sum_{ki,k'j} Q_{ij}(Z)\exp\left(i(\boldsymbol{k}-\boldsymbol{k}')\boldsymbol{R}_{0\|}\right)a^+_{ik}a_{j,k'},$$
$$Q_{ij}(Z) = \langle i|W(Z)|j\rangle, \quad i,j = 1(V_\uparrow),2(C_\uparrow),3(V_\downarrow),4(C_\downarrow), \tag{6.16}$$

where $a^+_{ik}(a_{j,k'})$ is the creation (annihilation) operator of a real fermion mode (6.15).

The matrix Q describes the *s-d* interaction of observable surface electrons with a magnetic atom. The interaction depends on the energy of the electron as well as on the position of the localized spin with respect to the surfaces. The approximate matrix Q at $k \to 0$ corresponds to low-energy electrons:

$$Q_{11} = Q_{44} = -Q_{22} = -Q_{33} = g_1 J_t(\boldsymbol{S}\times\boldsymbol{n})_z,$$
$$Q_{12} = Q_{21} = -Q_{34} = -Q_{43} = 2J_t\,|V|\,(\boldsymbol{S}\times\boldsymbol{n})_z,$$
$$Q_{13} = Q^*_{31} = (2VJ_{z2} - J_{z1})S_z - ig_1 J_t(\boldsymbol{Sn}),$$
$$Q_{14} = Q_{23} = Q^*_{41} = Q^*_{32} = -\text{sgn}\left[\tilde{V}_{AS}\right](J_{z2}\,g_1\,S_z + 2iV\,J_t(\boldsymbol{Sn})),$$
$$Q_{24} = Q^*_{42} = -(2VJ_{z2} + J_{z1})S_z + ig_1 J_t(\boldsymbol{Sn}),$$
$$g_1 = \frac{\Delta_S}{\sqrt{\Delta_S^2 + 4\tilde{V}^2_{as}}}, \quad V = \frac{\tilde{V}_{as}}{\sqrt{\Delta_S^2 + 4\tilde{V}^2_{as}}}, \quad \boldsymbol{n} = \boldsymbol{k}/k. \tag{6.17}$$

The interaction (6.17) is vertical voltage-controlled and depends on the *Z*-position of the magnetic atom inside the film through the constants *J*(*Z*) defined in (6.11). The spin structure of *s-d* interaction depends on the mutual orientations of the in-plane electron momentum and the vector of localized spin on which the electron is scattered. Some particular cases are given below.

1. An out-of-plane localized spin:

$$Q_{13} = Q^*_{31} = (2VJ_{z2} - J_{z1})S_z,$$
$$Q_{24} = Q^*_{42} = -(2VJ_{z2} + J_{z1})S_z,$$
$$Q_{14} = Q_{23} = Q^*_{41} = Q^*_{32} = -\text{sgn}\left[\tilde{V}_{AS}\right]J_{z2}\,g_1\,S_z. \tag{6.18}$$

2. An in-plane localized spin:

$$Q_{11} = Q_{44} = -Q_{22} = -Q_{33} = g_1 J_t(\boldsymbol{S}\times\boldsymbol{n})_z,$$
$$Q_{12} = Q_{21} = -Q_{34} = -Q_{43} = 2J_t\,|V|(\boldsymbol{S}\times\boldsymbol{n})_z,$$
$$Q_{13} = Q^*_{31} = -ig_1 J_t(\boldsymbol{Sn}),$$
$$Q_{14} = Q_{23} = Q^*_{41} = Q^*_{32} = -2iV\text{sgn}\left[\tilde{V}_{AS}\right]\,J_t(\boldsymbol{Sn}),$$
$$Q_{24} = Q^*_{42} = ig_1 J_t(\boldsymbol{Sn}). \tag{6.19}$$

3. A top-bottom vertically coupled symmetric film $\left(\Delta_S \neq 0,\ \tilde{V}_{as} \to 0\right)$:

$$Q_{11} = Q_{44} = -Q_{22} = -Q_{33} = J_t(\boldsymbol{S} \times \boldsymbol{n})_z,$$
$$Q_{13} = Q_{31}^* = -J_{z1}S_z - iJ_t(\boldsymbol{Sn}),$$
$$Q_{24} = Q_{42}^* = -J_{z1}S_z + iJ_t(\boldsymbol{Sn}). \tag{6.20}$$

4. A thick and vertically biased film $\left(\Delta_S,\ J_t \to 0,\ \tilde{V}_{as} \neq 0\right)$:

$$Q_{13} = Q_{31} = \left(\mathrm{sgn}\left[\tilde{V}_{AS}\right] J_{z2} - J_{z1}\right)S_z,$$
$$Q_{24} = Q_{42} = -\left(\mathrm{sgn}\left[\tilde{V}_{AS}\right] J_{z2} + J_{z1}\right)S_z. \tag{6.21}$$

The intra-band constants (6.21) describe the *s-d* interaction on independent surfaces as illustrated in Fig. 6.2.

The low-energy effective *s-d* interaction model has been formulated here by projecting the bulk *s-d* interaction onto surface states. Magnetic atoms interact with the surface electrons through the position-sensitive *s-d* interaction that can be controlled by a gate bias. The low-energy Hamiltonian (6.16) presents the position-dependent *s-d* interaction between the localized spin and real fermions. The explicit spin structure of the interaction depends on the electron momentum and spin mutual directions.

The *s-d* interaction, in turn, generates the indirect exchange between two magnetic atoms located somewhere in the TI film. The indirect exchange via gapless or gapped Dirac fermions would carry *Z*-dependence through the *s-d* coupling, so conditions for ferromagnetism would depend on the positions of the local spins with respect to surfaces. The indirect exchange interaction, which stems from *s-d* coupling (6.16), will be discussed in the next chapter.

References

1. Liu, Q., Liu, C.-X., Xu, C., Qi, X.-L., Zhang, S.-C.: Phys. Rev. Lett. **102**, 156603 (2009)
2. Zhu, J.-J., Yao, D.X., Zhang, S.-C., Chang, K.: Phys. Rev. Lett. **106**, 097201 (2011)
3. Abanin, D.A., Pesin, D.A.: Phys. Rev. Lett. **106**, 136802 (2011)
4. Checkelsky, J.G., Ye, J., Onose, Y., Iwasa, Y., Tokura, Y.: Nat. Phys. **8**, 729 (2012)
5. Ye, F., Ding, G.H., Zhai, H., Su, Z.B.: Europhys. Lett. **90**, 47001 (2011)
6. Sun, J., Chen, L., Lin, H.-Q.: Phys. Rev. **B89**, 115101 (2014)
7. Litvinov, V.I.: Phys. Rev. **B89**, 235316 (2014)
8. Zhu, Z.-H., Levy, G., Ludbrook, B., Veenstra, C.N., Rosen, J.A., Comin, R., Wong, D., Dosanjh, P., Ubaldini, A., Syers, P., Butch, N.P., Paglione, J., Efimov, I.S., Damascelli, A.: Phys. Rev. Lett. **107**, 186405 (2011)
9. Abrikosov, A.A.: Fundamentals of the Theory of Metals. Dover Publications, Mineola, NY (2017)
10. Kittel, C.: Quantum Theory of Solids. Wiley, New York (1987)
11. Ruderman, M.A., Kittel, C.: Phys. Rev. **96**, 99 (1954)
12. Kasuya, T.: Progr. Theor. Phys. **16**, 45 (1956)
13. Yosida, K.: Phys. Rev. **106**, 893 (1957)
14. Litvinov, V.I., Dugaev, V.K.: Phys. Rev. **B58**, 3584 (1996)
15. Litvinov, V.I.: Wide Bandgap Semiconductor Spintronics. Pan Stanford, Singapore (2016)

Chapter 7
Indirect Exchange Interaction Mediated by Dirac Fermions

The indirect exchange interaction carries the signature of the topological surface states. The most direct illustration of the relation between the topological order and the magnetic exchange can be seen in a sample where the Fermi level is tuned to the Dirac point: the magnetic field-induced violation of time reversal symmetry destroys the topological protection and turns massless fermions into massive ones. This topological phase transition drastically changes the character of the exchange range function: from one mediated by gapless fermions and scaled as R^{-2} [1] to that mediated by massive excitations and then demonstrating the monotonic and short-ranged exponential character typical of an intrinsic semiconductor, the Bloembergen-Rowland (BR) interaction [2].

The spin–electron interaction, discussed in Chap. 6, serves as a starting point in calculations of the indirect exchange interaction between magnetic atoms mediated by surface excitations. Below we consider two typical settings where surface electrons are either degenerate, providing for a well-defined Fermi surface, or non-degenerate, with the chemical potential located in the top-bottom tunneling energy gap.

7.1 Generic Indirect Exchange via Conduction Electrons

The *s-d* interaction (6.3) can be extended for a two-band solid as the matrix generalization of single-band expression:

V. Litvinov, *Magnetism in Topological Insulators*,
https://doi.org/10.1007/978-3-030-12053-5_7

$$H = H_{\rm el} + H_{\rm sd}, \quad H_{\rm el} = \sum_{\boldsymbol{k}} a^+_{\boldsymbol{k}} [\hat{\epsilon}(k) \otimes I] a_{\boldsymbol{k}},$$
$$H_{\rm sd} = \frac{1}{N} \sum_{\boldsymbol{kq}} \sum_{i} e^{i\boldsymbol{qR}_i} a^+_{\boldsymbol{k}} \left[\hat{J} \otimes \boldsymbol{\sigma S}_i\right] a_{\boldsymbol{k}+\boldsymbol{q}}, \tag{7.1}$$

where $a_{\boldsymbol{k}}$ is the 4-spinor in bands and spins, $\hat{J}_{jl} = Ju^*_j(\boldsymbol{R}_i)u_l(\boldsymbol{R}_i)$ is the 2×2 matrix operating in the space of two bands (j, l = 1, 2). The unit matrix I and the spin matrices $\boldsymbol{\sigma}$ act in the 2×2 electron spin space. Since the matrix products in (7.1) are comprised of matrices acting in different spaces, the product is understood as the Kronecker product.

The indirect exchange interaction between two localized spins appears as a second-order energy correction with respect to $H_{\rm sd}$. It is convenient to express the second-order correction by the diagram shown in Fig. 7.1:

The diagram depicts the process in which an impurity spin creates the virtual electron-hole pair (solid lines) which then propagates to another impurity and annihilates there. Two localized spins exchange the virtual electron-hole pair, maintaining the indirect exchange interaction between them. The process, shown in Fig. 7.1, can be called vacuum polarization if the vacuum is understood to be either a Fermi sphere in a single-band metal (RKKY model), or fully occupied valence band and empty conduction band in an intrinsic semiconductor at zero absolute temperature.

In a two-band solid,

$$\hat{\epsilon}(k) = \begin{pmatrix} \epsilon_1 & 0 \\ 0 & \epsilon_2 \end{pmatrix}, \quad \hat{J} = \begin{pmatrix} J_1 & J_{12} \\ J_{12} & J_2 \end{pmatrix},$$
$$\epsilon_{1,2} = \pm(\hbar^2 k^2/2m_{\rm c,v} + E_g/2) \tag{7.2}$$

where $E_{\rm g}$ is the bandgap, the reference energy is chosen in the middle of the bandgap.

The propagation of electrons and holes can be described by the Green function. In the finite temperature Matsubara technique, the Green function is expressed as [3]:

$$G(k, \omega_m) = [(i\omega_m + \mu) - H_{el}]^{-1} = \begin{pmatrix} \dfrac{I}{i\omega_m + \mu - \epsilon_1} & 0 \\ 0 & \dfrac{I}{i\omega_m + \mu - \epsilon_2} \end{pmatrix}, \tag{7.3}$$

where $\omega_m = (2n + 1)\pi T$ is the Matsubara frequency, $n = 0, \pm 1, \ldots$, T is the temperature in energy units, and μ is the chemical potential. Each arrow in Fig. 7.1 corresponds to the electron Green function and each vertex contains the *s-d* coupling

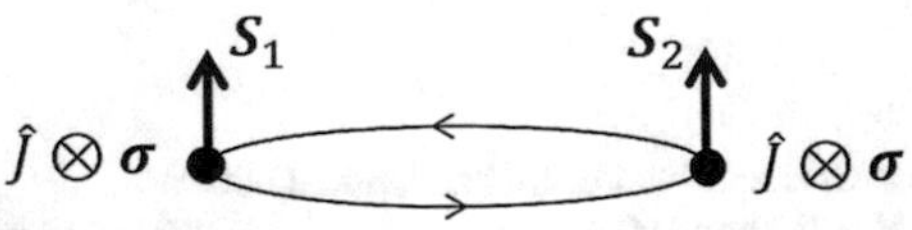

Fig. 7.1 Indirect exchange interaction

matrix $\hat{J} \otimes \boldsymbol{\sigma} \exp(i\boldsymbol{q}\boldsymbol{R}_{1,2})$. The analytical expression follows from the rules of diagrammatic technique:

$$H = \frac{T}{N^2} \sum_{n=-\infty}^{\infty} \sum_{\boldsymbol{k},\boldsymbol{q}} \exp(i\boldsymbol{q}\boldsymbol{R}) Tr\{\hat{J} \otimes \boldsymbol{\sigma}\boldsymbol{S}_1 G(\boldsymbol{k},\omega_n)\hat{J} \otimes \boldsymbol{\sigma}\boldsymbol{S}_2 G(\boldsymbol{k}+\boldsymbol{q},\omega_n)\},$$
$$\boldsymbol{R} = \boldsymbol{R}_1 - \boldsymbol{R}_2. \tag{7.4}$$

The expression (7.4) is written in generic form, where the dimension of the matrices depends on the type of electron spectrum in the host. The operation $Tr = Tr_s Tr_j$ implies summation over the diagonal matrix elements that is summation over electron quantum numbers: bands (j) and spins (s). In the model (7.2) and (7.3), after trace over band indexes we get

$$H = \frac{T}{N^2} \sum_{n=-\infty}^{\infty} \sum_{\boldsymbol{k},\boldsymbol{q}} \exp(i\boldsymbol{q}\boldsymbol{R}) \left[J_1^2 G_{11}(\boldsymbol{k},\omega_n) G_{11}(\boldsymbol{k}+\boldsymbol{q},\omega_n) \right.$$
$$\left. + J_2^2 G_{22}(\boldsymbol{k},\omega_n) G_{22}(\boldsymbol{k}+\boldsymbol{q},\omega_n) + 2|J_{12}|^2 G_{11}(\boldsymbol{k},\omega_n) G_{22}(\boldsymbol{k}+\boldsymbol{q},\omega_n) \right] Tr_s\{(\boldsymbol{\sigma}\boldsymbol{S}_1)(\boldsymbol{\sigma}\boldsymbol{S}_2)\} \tag{7.5}$$

and, after summation over spin indexes, $Tr_s\{(\boldsymbol{\sigma}\boldsymbol{S}_1)(\boldsymbol{\sigma}\boldsymbol{S}_2)\} = \boldsymbol{S}_1\boldsymbol{S}_2$ we obtain the indirect exchange interaction in the form:

$$H = V(\boldsymbol{R})\boldsymbol{S}_1\boldsymbol{S}_2,$$
$$V(\boldsymbol{R}) = \frac{T}{N^2} \sum_{n=-\infty}^{\infty} \sum_{\boldsymbol{k},\boldsymbol{q}} \left[J_1^2 G_{11}(\boldsymbol{k},\omega_n) G_{11}(\boldsymbol{k}+\boldsymbol{q},\omega_n) + J_{22}^2 G_2(\boldsymbol{k},\omega_n) G_{22}(\boldsymbol{k}+\boldsymbol{q},\omega_n) \right.$$
$$\left. + 2|J_{12}|^2 G_{11}(\boldsymbol{k},\omega_n) G_{22}(\boldsymbol{k}+\boldsymbol{q},\omega_n) \right] \exp(i\boldsymbol{q}\boldsymbol{R}). \tag{7.6}$$

The calculation of the frequency sums can be done using the relation

$$T \sum_n g(i\omega_n) = - \sum_i \mathrm{Res}[g(x_i) f(x_i)],$$
$$f(x) = \frac{1}{2} tanh(x/2T), \tag{7.7}$$

where residues are taken in the poles of $g(x)$. At $T \to 0$, one may go to frequency integration by the substitution

$$i\omega_n \to \omega + i\delta \ sign(\omega), \quad T \sum_n \to \frac{1}{2\pi i} \int_{-\infty}^{\infty} d\omega, \quad \delta \to 0. \tag{7.8}$$

The frequency integrals in (7.6) contain terms

$$A_i = \int_\Gamma \frac{d\omega}{(\omega+\mu-\epsilon_i(k))(\omega+\mu-\epsilon_i(k+q))},$$
$$A_{12} = \int_\Gamma \frac{d\omega}{(\omega+\mu-\epsilon_1(k))(\omega+\mu-\epsilon_2(k+q))}, \tag{7.9}$$

which can be calculated by closing the integration path Γ in the upper or lower frequency half-plane. At this point, it is important to locate the positions of the poles. The actual situation depends on the energy spectrum of the host. Below we consider several options.

7.2 Three-Dimensional Metal: RKKY Model

In a simple metal, we deal with a single band, μ is the Fermi energy. The only relevant term in (7.6) is that proportional to J_1^2. The integration path is shown in Fig. 7.2. If the poles of the integrand in A_1, (7.9), are both positive or negative, one can close the integration path so that no poles lie inside the closed path and then the integral is equal to zero. This means that the nonzero A_1 is possible only if $p_1 = \varepsilon(\boldsymbol{k}) - \mu < 0$ and $p_2 = \varepsilon(\boldsymbol{k} + \boldsymbol{q}) - \mu > 0$, or vice versa. This corresponds to virtual electron-hole pair excitation across the Fermi level.

Integral A_1 can be calculated by closing the integration path in the upper frequency half-plane as shown in Fig. 7.2,

$$A_1 = \frac{1}{2\pi i}\int_\Gamma \frac{\exp(i\omega\tau)d\omega}{(\omega+\mu-\epsilon(\boldsymbol{k}))(\omega+\mu-\epsilon(\boldsymbol{k}+\boldsymbol{q}))}\bigg|_{\tau\to 0}$$
$$= \frac{1}{\epsilon(\boldsymbol{k})-\epsilon(\boldsymbol{k}+\boldsymbol{q})}. \tag{7.10}$$

It is instructive to note that the same result follows from relation (7.7) in the limit $T \to 0$:

$$T\sum_m \frac{1}{(i\omega_m+\mu-\epsilon(\boldsymbol{k}))(i\omega_m+\mu-\epsilon(\boldsymbol{k}+\boldsymbol{q}))}$$
$$= \frac{f(\epsilon(\boldsymbol{k})-\mu) - f(\epsilon(\boldsymbol{k}+\boldsymbol{q})-\mu)}{\epsilon(\boldsymbol{k})-\epsilon(\boldsymbol{k}+\boldsymbol{q})} \to \frac{1}{\epsilon(\boldsymbol{k})-\epsilon(\boldsymbol{k}+\boldsymbol{q})}. \tag{7.11}$$

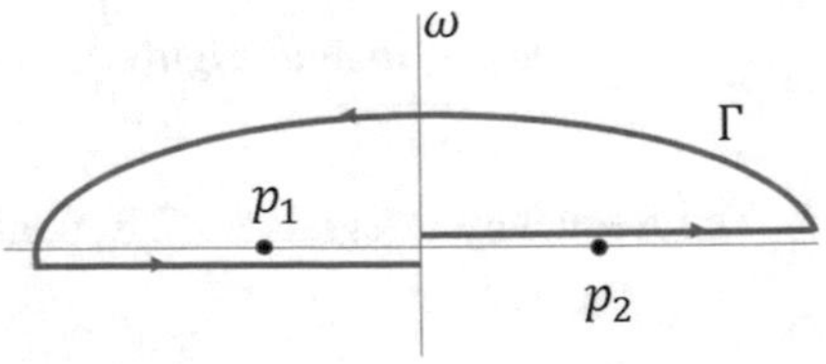

Fig. 7.2 Integration path and poles in integral A_1

The sum (7.11) is equal to zero if both the initial and the final scattering energies, $\epsilon(\boldsymbol{k})$ and $\epsilon(\boldsymbol{k}+\boldsymbol{q})$, happen to be on the same side with respect to the Fermi level, that is equivalent to our previous notion of zero integral if the poles are of the same sign in the frequency integral (7.10). After integration over frequency, the range function (7.6) is expressed as

$$V_{\mathrm{RKKY}}(\boldsymbol{R}) = \frac{J_1^2}{n^2\,(2\pi)^6}\int dk dk' \frac{\cos\left[\boldsymbol{R}(\boldsymbol{k}-\boldsymbol{k}')\right]}{\epsilon(\boldsymbol{k})-\epsilon(\boldsymbol{k}')}. \quad (7.12)$$

Integration limits in wave vectors are determined by the conditions that the initial wave vectors are inside the Fermi sphere, $k \le k_{\mathrm{F}}$, and the momentum transfer due to scattering on a localized spin, q, spans the interval $(0, \infty)$. The calculation of the integrals in Fig. 7.12 yields the RKKY result [4–6]:

$$\begin{aligned}
V_{\mathrm{RKKY}}(R) &= \frac{J_1^2}{n^2(2\pi)^6}\int_0^{2\pi} d\varphi\, d\varphi' \int \frac{k\prime^2 k^2 dk\, dk'}{\epsilon(k)-\epsilon(k')}\int_{-1}^{1} \cos(Rkx)\cos(Rk'x')dxdx' \\
&= \frac{mJ_1^2}{4\pi^4 n^2\hbar^2 R^2}\int_0^{K_{\mathrm{F}}} kdk \int_0^{\infty} \frac{\sin(Rk)\sin(Rk')}{k^2-k\prime^2}k'dk' \\
&= \frac{mJ_1^2}{8\, n^2\pi^3\hbar^2 R^2}\int_0^{k_{\mathrm{F}}} \cos(Rk)\sin(Rk)\, kdk \\
&= \frac{mJ_1^2}{8\, n^2\pi^3\hbar^2 R^2}\left[-\frac{k_{\mathrm{F}}\cos(2k_{\mathrm{F}}R)}{4R}+\frac{\sin(2k_{\mathrm{F}}R)}{8R^2}\right].
\end{aligned} \quad (7.13)$$

In (7.13), we used the relation

$$\int_0^{\infty} \frac{\sin(Rk')}{k^2-k\prime^2}k'dk' = \frac{\pi}{2}\cos(Rk). \quad (7.14)$$

As we assumed the energy spectrum parabolic that is valid for small wave vectors $k \ll 1/a$, a is the lattice constant, one can use the range function (7.14) in the range $R \gg a$. At large distances, $Rk_{\mathrm{F}} \gg 1$, $V_{\mathrm{RKKY}}(R)$ decays as $1/R^3$ and oscillaltes with the spatial period π/k_{F}. If the carrier density (electrons or holes) n_{c} is low, we deal the with a short-distance regime, $Rk_{\mathrm{F}} \ll 1$. The interaction is negative, so it favors the ferromagnetic ordering and scales as $k_{\mathrm{F}} \sim n_{\mathrm{c}}^{1/3}$.

In an impure metal, the electron scattering can be taken into account by introducing a damping term in (7.10): $\omega \to \omega + i\hbar/\tau$, τ is the lifetime with respect to scattering on non-magnetic centers. As a result, the RKKY range function acquires the additional decay factor [7, 8]:

$$\tilde{V}(R) = V_{\mathrm{RKKY}}(R)\exp(-R/l), \quad (7.15)$$

where $l = \hbar k_{\mathrm{F}}\tau/m$ is the electron mean free path.

7.3 Exchange Interaction in Semiconductors

In a semiconductor, the chemical potential lies in the bandgap, $\mu = 0$, and there are no carriers in the ground state, $T = 0$. The poles in the integrand of A_1 or A_2 have the same signs, and then $A_1 = A_2 = 0$. The relevant term in (7.6) is proportional to $|J_{12}|^2 A_{12}$. That means the exchange interaction is caused by the inter-band *s-d* coupling that generates virtual electron-hole pairs excited across the bandgap. The pairs excited at one localized spin propagate through the crystal until they annihilate at another localized spin. That is the case for Bloembergen-Rowland interaction [9]. Using the energy spectrum from (7.2), one gets

$$V_{\mathrm{BR}}(R) = \frac{J_{12}^2}{N^2} \sum_{\boldsymbol{k},\boldsymbol{q}} \frac{1}{2\pi i} \int_\Gamma d\omega \frac{\exp(i\boldsymbol{q}\boldsymbol{R})}{(\omega - \epsilon_1(\boldsymbol{k}))(\omega - \epsilon_2(\boldsymbol{k}+\boldsymbol{q}))}$$

$$= -\frac{m J_{12}^2}{32 n^2 \hbar^2 \pi^6} \int dk dk' \frac{\cos\left[\boldsymbol{R}(\boldsymbol{k} - \boldsymbol{k}')\right]}{k^2 + k'^2 + \lambda^2}, \quad \lambda = \sqrt{\frac{2mE_{\mathrm{g}}}{\hbar^2}}. \tag{7.16}$$

Unlike the one-band RKKY in metals, the momentum integration in (7.16) does not have restrictions:

$$V_{\mathrm{BR}}(R) = -\frac{m J_{12}^2}{4\pi^4 n^2 \hbar^2 R^2} \int_0^\infty k dk \sin(Rk) \int_{-\infty}^{\infty} \frac{k' \sin(Rk')}{k^2 + k'^2 + \lambda^2} dk'$$

$$= \frac{i\, m J_{12}^2}{4\pi^4 n^2 \hbar^2 R^2} \int_0^\infty k dk \sin(Rk) \int_{\Gamma_1} \left[\frac{k' \exp(iRk')}{k^2 + k'^2 + \lambda^2}\right] dk'. \tag{7.17}$$

Contour Γ_1 is shown in Fig. 7.3, $p_{1,2} = \pm i\sqrt{k^2 + \lambda^2}$. The negative sign in (7.17) comes from the difference of distribution functions in (7.11), implying that at $T \to 0$ the conduction band is empty, $f(\epsilon_1 - \mu) \to 0$, and the valence band is fully occupied, $f(\epsilon_2 - \mu) \to 1$.

Integration over k' gives $2\pi i$ multiplied by the residue in the pole p_1:

$$V_{\mathrm{BR}}(R) = -\frac{m J_{12}^2}{8\pi^3 n^2 \hbar^2 R^4} I,$$

$$I = \int_0^\infty x \sin(x) \exp\left(-\sqrt{x^2 + x_0^2}\right) dx,$$

$$x_0 = R/R_0, \quad R_0 = \hbar\left(2mE_{\mathrm{g}}\right)^{-1/2}. \tag{7.18}$$

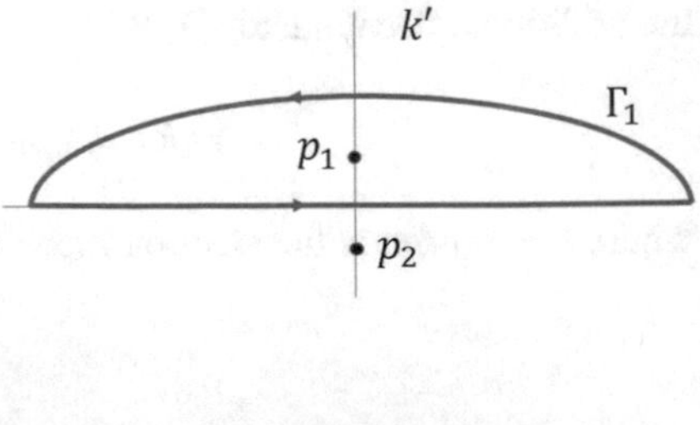

Fig. 7.3 Integration contour in (7.17)

In order to evaluate the integral, we take the derivative $\partial I/\partial x_0$ and find

$$\frac{\partial I}{\partial x_0} = -\frac{x_0^2}{\sqrt{2}} K_1\left(x_0\sqrt{2}\right), \tag{7.19}$$

where $K_\nu(z)$ is the modified Bessel function. Making use of the relation $\partial[z^\nu K_\nu(z)]/\partial z = -z^\nu K_{\nu-1}(z)$, one gets $I = \frac{1}{2}x_0^2 K_2\left(\sqrt{2}x_0\right)$ and, finally, the interaction takes the form [10]:

$$V_{\mathrm{BR}}(R) = -\frac{mJ_{12}^2}{16\,\pi^3 n^2 \hbar^2 R^4}\begin{cases} 1, & R \ll R_0 \\ \frac{\sqrt{\pi}}{2^{3/4}}\left(\frac{R}{R_0}\right)^{3/2} \exp\left(-\frac{\sqrt{2}R}{R_0}\right), & R \gg R_0 \end{cases}. \tag{7.20}$$

In the long-range limit, the interaction (7.20) reproduces the one calculated in [11], while in the short-range limit it scales as $1/R^4$ [12]. The exponential decay of the range function is determined by the energy gap and effective mass. We have considered a simplistic model which keeps the effective masses of electrons and holes equal. In a more realistic approach, the effective mass in (7.20) should be replaced with a reduced mass: $m \to m_c m_v/(m_c + m_v)$.

Since there are no real carriers to mediate the indirect exchange in a semiconductor at $T = 0$ the Bloembergen-Rowland interaction may be referred to as a lattice contribution to the magnetic interaction in dielectrics. Interaction $V_{\mathrm{BR}}(R)$ does not contain the oscillations obtained in the original paper [9]. These oscillations appear as a result of cutting the integral over momentum on the upper limit, and as pointed out in [11], they cannot be considered to be real.

7.4 Exchange Interaction via Surface Fermions

In Chap. 6, we discussed the *s-d* interaction in the representation which diagonalizes the electron Hamiltonian (6.15), so the Green function is a diagonal matrix and, in the finite temperature Matsubara technique, can be expressed as:

$$\begin{gathered} G(\boldsymbol{k},\omega_m) = \mathrm{diag}\left[(i\omega_n - E_{\mathrm{v}\uparrow} + \mu)^{-1}, (i\omega_n - E_{\mathrm{c}\uparrow} + \mu)^{-1}, (i\omega_n - E_{\mathrm{v}\downarrow} + \mu)^{-1}, (i\omega_n - E_{\mathrm{c}\downarrow} + \mu)^{-1}\right], \\ E_{\mathrm{c,v}\uparrow} = Dk^2 \pm R_\uparrow, \quad E_{\mathrm{c,v}\downarrow} = Dk^2 \pm R_\downarrow, \\ R_{\uparrow\downarrow} = \sqrt{\left(\tfrac{\Delta_S}{2} - Bk^2\right)^2 + \left(\tilde{A}_2 k \pm \tilde{V}_{AS}\right)^2}, \end{gathered} \tag{7.21}$$

Each arrow in Fig. 7.1 corresponds to an electron Green function and each vertex contains 4×4 matrix Q from (6.16) and the factor $\exp(i\boldsymbol{q}\boldsymbol{R}_{1,2})$. The indirect exchange interaction between two magnetic atoms separated by the vector $\boldsymbol{R}_{\|} = \boldsymbol{R}_{\|1} - \boldsymbol{R}_{\|2}$ can be expressed as

$$H(Z_1, Z_2) = \frac{T}{2(2\pi)^4 n^2} \sum_{\omega_n} \iint dk\, dk'$$
$$\exp\left[iR_{\|}(k - k')\right] Tr\{Q(Z_1, n) G(k, \omega_n) Q(Z_2, n') G(k', \omega_n)\}. \tag{7.22}$$

The trace operation runs over four quantum numbers: bands and spins. The structure of interaction (7.22) depends on the position of the chemical potential.

In a slab with degenerate surface electrons ($\mu > \sqrt{\Delta_s^2/4 + \tilde{V}_{as}^2}$), the leading interaction term is of the RKKY-type which originates from excitations around the Fermi energy in the surface conduction band $\varepsilon_{\mathrm{F}} = \mu - \sqrt{\Delta_s^2/4 + \tilde{V}_{as}^2}$. The intra-band contribution to indirect exchange (7.22) is expressed as

$$\begin{aligned} H_{cc} = & \frac{T}{2(2\pi)^4 n^2} \sum_{\omega_n} \iint dk\, dk' \exp\left[iR_{\|}(k - k')\right] \\ & \times \{g_1^2 M(Z_1, Z_2) \left[G_{c\uparrow}(k, \omega_n) G_{c\uparrow}(k', \omega_n) + G_{c\downarrow}(k, \omega_n) G_{c\downarrow}(k', \omega_n)\right] \times (S_1 \times n)_z (S_2 \times n)_z \\ & + 2 g_1^2 M(Z_1, Z_2)\, G_{c\uparrow}(k, \omega_n) G_{c\downarrow}(k', \omega_n)(S_1 n)(S_2 n') \\ & + 2L(Z_1, Z_2)\, G_{c\uparrow}(k, \omega_n) G_{c\downarrow}(k', \omega_n) S_{1z} S_{2z}\}, \end{aligned} \tag{7.23}$$

where

$$\begin{aligned} M(Z_1, Z_2) &= J_t(Z_1) J_t(Z_2), \\ L(Z_1, Z_2) &= \left[J_{z1}(Z_1) + 2V J_{z2}(Z_1)\right]\left[J_{z1}(Z_2) + 2V J_{z2}(Z_2)\right], \\ g_1 &= \frac{\Delta_S}{\sqrt{\Delta_S^2 + 4\tilde{V}_{as}^2}}, \quad V = \frac{\tilde{V}_{as}}{\sqrt{\Delta_S^2 + 4\tilde{V}_{as}^2}}. \end{aligned} \tag{7.24}$$

If the chemical potential is located in the energy gap of the surface spectrum, $\mu < \sqrt{\Delta_s^2/4 + \tilde{V}_{as}^2}$, there are no carriers in the surface bands at $T = 0$, and the indirect exchange stems from inter-band excitations across the gap:

$$\begin{aligned} H_{cv} = & \frac{T}{2(2\pi)^4 n^2} \sum_{\omega_n} \iint dk\, dk' \exp\left[iR_{\|}(k - k')\right] \\ & \times \{8V^2 M(Z_1, Z_2) \left[G_{v\uparrow}(k, \omega_n) G_{c\uparrow}(k', \omega_n) + G_{v\downarrow}(k, \omega_n) G_{c\downarrow}(k', \omega_n)\right] (S_1 \times n)_z (S_2 \times n)_z \\ & + 8V^2 M(Z_1, Z_2) \left[G_{c\uparrow}(k, \omega_n) G_{v\downarrow}(k', \omega_n) + G_{v\uparrow}(k, \omega_n) G_{c\downarrow}(k', \omega_n)\right] (S_1 n)(S_2 n') \\ & + 2 g_1^2 K(Z_1, Z_2) \left[G_{c\uparrow}(k, \omega_n) G_{v\downarrow}(k', \omega_n) + G_{v\uparrow}(k, \omega_n) G_{c\downarrow}(k', \omega_n)\right] S_{1z} S_{2z}\}, \\ & K(Z_1, Z_2) = J_{z2}(Z_1) J_{z2}(Z_2). \end{aligned} \tag{7.25}$$

To find the spin texture of the indirect exchange, we perform integration over angles in (7.23) and (7.25):

$$
\begin{aligned}
I_{1\varphi} &= S_{1z}S_{2z}\int \exp\left[i\boldsymbol{R}_{\parallel}(\boldsymbol{k}-\boldsymbol{k}')\right]d\varphi_1 d\varphi_2 \\
&= \int_0^{2\pi} d\varphi_1 \exp\left[ikR_{\parallel}\cos(\varphi_1)\right]\int_0^{2\pi} d\varphi_2 \exp\left[-ik'R_{\parallel}\cos(\varphi_2)\right] \\
&= 4\pi^2 J_0(kR_{\parallel})J_0(k'R_{\parallel})S_{1z}S_{2z},
\end{aligned}
\tag{7.26}
$$

$$
\begin{aligned}
I_{2\varphi} &= \int (\boldsymbol{S_1 n})(\boldsymbol{S_2 n}')\exp\left[i\boldsymbol{R}_{\parallel}(\boldsymbol{k}-\boldsymbol{k}')\right]d\varphi_1 d\varphi_2 \\
&= S_1S_2\int_0^{2\pi} d\varphi_1 \cos(\alpha_1)\exp\left[ikR_{\parallel}\cos(\varphi_1)\right]\int_0^{2\pi} d\varphi_2 \cos(\alpha_2)\exp\left[-ik'R_{\parallel}\cos(\varphi_2)\right] \\
&= \frac{4\pi^2}{R_{\parallel}^2}J_1(kR_{\parallel})J_1(k'R_{\parallel})(\boldsymbol{S_1R}_{\parallel})(\boldsymbol{S_2R}_{\parallel}),
\end{aligned}
\tag{7.27}
$$

$$
\begin{aligned}
I_{3\varphi} &= \int (\boldsymbol{S_1}\times\boldsymbol{n})_z(\boldsymbol{S_2}\times\boldsymbol{n}')_z \exp\left[i\boldsymbol{R}_{\parallel}(\boldsymbol{k}-\boldsymbol{k}')\right]d\varphi_1 d\varphi_2 \\
&= \frac{4\pi^2}{R_{\parallel}^2}J_1(kR_{\parallel})J_1(k'R_{\parallel})(\boldsymbol{S_1}\times\boldsymbol{R}_{\parallel})_z(\boldsymbol{S_2}\times\boldsymbol{R}_{\parallel})_z,
\end{aligned}
\tag{7.28}
$$

where $J_{0,1}(x)$ is the Bessel function of the first kind. Calculations were performed using $\alpha_{1,2}$ expressed in terms of angles $\gamma_{1,\,2}$ between spin vectors and the vector $\boldsymbol{R}_{\parallel}$ that connects them, as illustrated in Fig. 7.4:

$$
\begin{aligned}
\cos(\alpha_1) &= \cos(\gamma_1)\cos(\varphi_1) - \sin(\gamma_1)\sin(\varphi_1) \\
\cos(\alpha_2) &= \cos(\gamma_2)\cos(\varphi_2) - \sin(\gamma_2)\sin(\varphi_2).
\end{aligned}
\tag{7.29}
$$

Using (7.26)–(7.29), the intra-band contribution (7.23) can be written as

$$
\begin{aligned}
H_{cc} = \frac{1}{8\pi^2\,n^2}\Bigg\{&\frac{1}{R_{\parallel}^2}g_1^2 M(Z_1,Z_2)\left[F^1_{c\uparrow c\uparrow}+F^1_{c\downarrow c\downarrow}\right](\boldsymbol{S_1}\times\boldsymbol{R}_{\parallel})_z(\boldsymbol{S_2}\times\boldsymbol{R}_{\parallel})_z \\
&+\frac{2}{R_{\parallel}^2}g_1^2 M(Z_1,Z_2)F^1_{c\uparrow c\downarrow}(\boldsymbol{S_1R}_{\parallel})(\boldsymbol{S_2R}_{\parallel}) + 2L(Z_1,Z_2)F^0_{c\uparrow c\downarrow}\,S_{1z}S_{2z}\Bigg\},
\end{aligned}
\tag{7.30}
$$

where

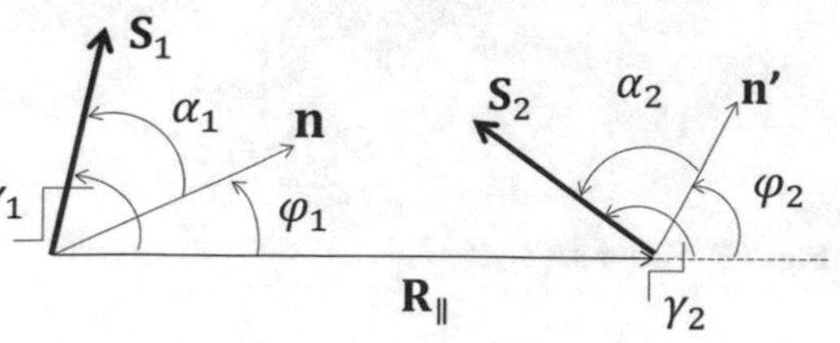

Fig. 7.4 Angles in integrals (7.26) and (7.27)

$$
\begin{aligned}
F^{0,1}_{isjs'} &= T\sum_{\omega_n} G^{0,1}_{is}\left(\omega_n, R_\parallel\right)G^{0,1}_{js'}\left(\omega_n, R_\parallel\right), \quad i,j = (c, \mathrm{v});\ s, s' = (\uparrow, \downarrow), \\
G^{0,1}_{is}\left(\omega_n, R_\parallel\right) &= \int_0^\infty kdk\, J_{0,1}\left(kR_\parallel\right)\left(i\omega_n - E_{is}(k) + \mu\right)^{-1}, \quad R_\parallel = \left|\boldsymbol{R}_\parallel\right|.
\end{aligned} \tag{7.31}
$$

The spin structure of the RKKY-type indirect exchange (7.30) is consistent with that obtained in [13].

If the Fermi level is in the top-bottom mixing gap, the only contribution to the indirect exchange (7.25) comprises inter-band terms

$$
\begin{aligned}
H_{c\mathrm{v}} = \frac{1}{4\pi^2 n^2} \times \Big\{ &\frac{4}{R_\parallel^2} V^2 M(Z_1, Z_2)\left[F^1_{\mathrm{v}\uparrow c\uparrow} + F^1_{\mathrm{v}\downarrow c\downarrow}\right]\left(\boldsymbol{S}_1 \times \boldsymbol{R}_\parallel\right)_z\left(\boldsymbol{S}_2 \times \boldsymbol{R}_\parallel\right)_z \\
&+ \frac{4}{R_\parallel^2} V^2 M(Z_1, Z_2)\ \left[F^1_{c\uparrow \mathrm{v}\downarrow} + F^1_{\mathrm{v}\uparrow c\downarrow}\right]\ \left(\boldsymbol{S}_1\boldsymbol{R}_\parallel\right)\left(\boldsymbol{S}_2\boldsymbol{R}_\parallel\right) \\
&+ g_1^2\ K(Z_1, Z_2)\left[F^0_{c\uparrow \mathrm{v}\downarrow} + F^0_{\mathrm{v}\uparrow c\downarrow}\right] S_{1z}S_{2z}\Big\}.
\end{aligned} \tag{7.32}
$$

Some conclusions on possible magnetic phases can be drawn without the actual calculation of range functions. Coefficients K, L, and M carry the dependence of indirect exchange on the positions of magnetic impurities along the z-direction.

The coefficient $M(Z_1, Z_2) = J_t(Z_1)J_t(Z_2)$ in (7.30) and (7.32) is the product of two factors which both tend to zero in the vicinity of either of the surfaces, as shown in Fig. 7.5. Coordinates $Z = -20\,\dot{A}$ and $Z = 20\,\dot{A}$ correspond to the top and bottom surfaces, respectively.

So, if magnetic atoms are located on either of the two surfaces, the terms proportional to M, since they are the square of the top-bottom overlap integral, are negligible and the interaction is of the Ising-type spin texture favoring magnetic ordering perpendicular to the film surfaces (last terms in (7.30) and (7.32)). The M-terms come into play if the magnetic atoms are close to the middle of the film in the z-direction, and the magnetic ordering caused by M-terms has a complicated spin

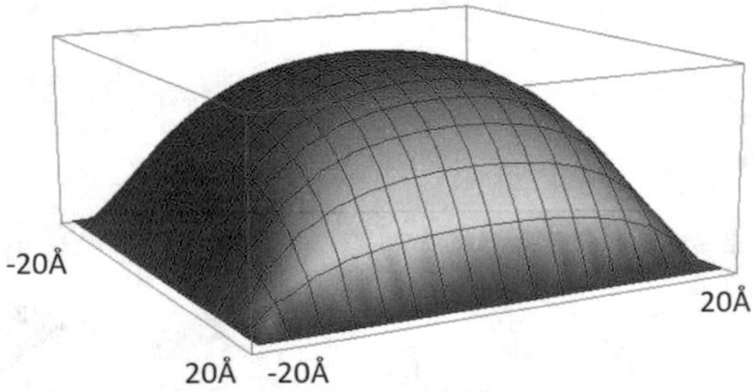

Fig. 7.5 Factor $M(Z_1, Z_2)/J_\parallel^2$

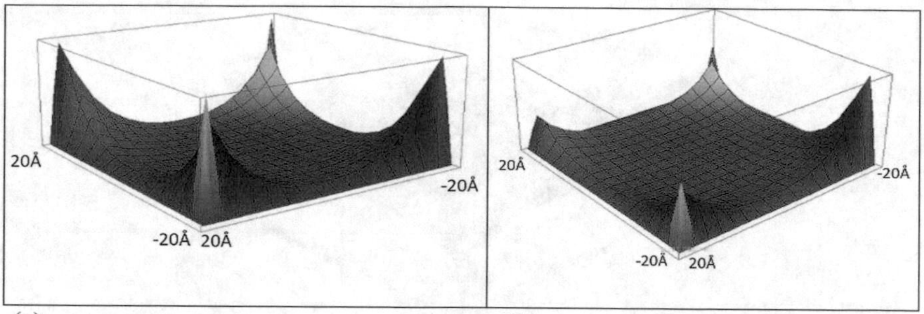

(a) (b)

Fig. 7.6 Coefficient $L(Z_1, Z_2)/J_z^2$ at different gate voltages. (**a**) $V_g = 0$. (**b**) $V_g = -12$ mV

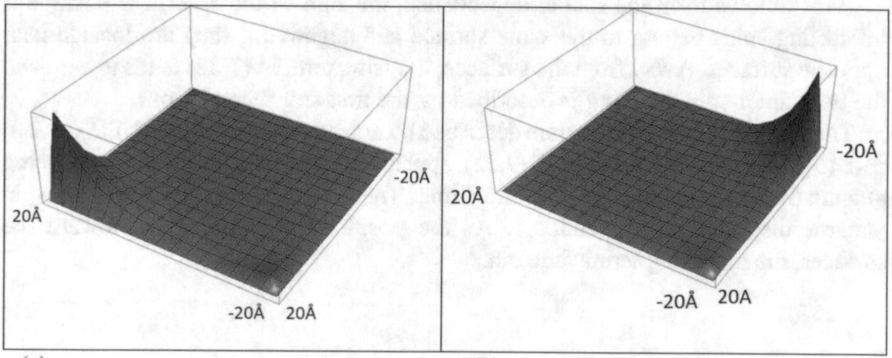

(a) (b)

Fig. 7.7 Coefficient $L(Z_1, Z_2)/J_z^2$. (**a**) $V_g = -120$ mV. (**b**) $V_g = 120$ mV

pattern depending on the angles between the vector connecting two impurities and their spin vectors.

The coefficient $L(Z_1, Z_2)$ determines the vertical voltage-controlled spatial dependence of the Ising term in (7.30) as illustrated in Figs. 7.6 and 7.7.

In an unbiased film (Fig. 7.6a), the Ising-type term is insensitive to localized spin positions on any of the two surfaces and the indirect exchange is controlled by the range factor as a function of the in-plane distance between spins. An applied vertical voltage affects this part of the interaction, gradually enhancing it on one of the surfaces and suppressing it on the other one, as shown in Fig. 7.6b and also in Fig. 7.7 where high enough voltages of opposite signs switch the Ising term between the two opposite surfaces.

For the magnetic atoms located near the surfaces, the inter-band indirect exchange (7.32) is fully determined by the K-term (7.25) whose sign depends on the mutual positions of interacting magnetic atoms as rendered in Fig. 7.8.

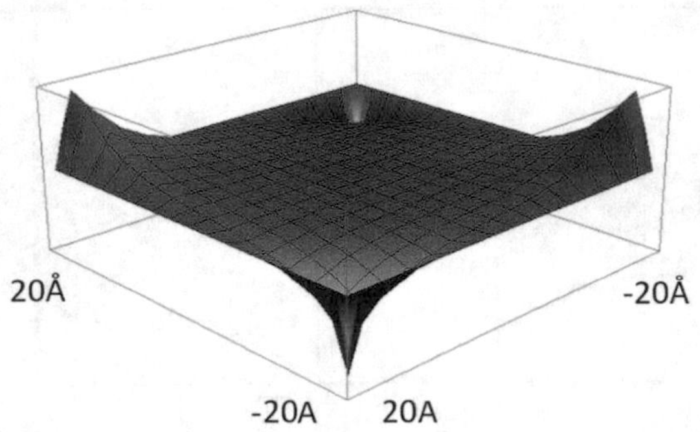

Fig. 7.8 Inter-band exchange parameter $K(Z_1, Z_2)/J_z^2$

As it follows from the spatial dependence, the sign of the *K*-term is positive if interacting spins belong to the same surface and negative if they are located near opposite surfaces. Away from the surfaces, the Ising term in (7.32) tends to zero and the weak inter-spin coupling is described by the first and second terms.

The complex planar spin texture described by non-Ising terms $(\mathbf{S_1} \times \boldsymbol{R}_{\|})_z(\mathbf{S_2} \times \boldsymbol{R}_{\|})_z$ and $(\mathbf{S_1}\boldsymbol{R}_{\|})(\mathbf{S_2}\boldsymbol{R}_{\|})$ in (7.30) and (7.32) appears in the middle of the slab being quantitatively significant only in thin films. These terms favor the localized spin patterns discussed in Appendix 1. As the positions of spins move toward the surfaces, the non-Ising terms fade out.

7.5 RKKY Range Function in Inversion Asymmetric TI Films

Since it is not easy to pin the chemical potential to the middle of the surface energy gap, the most probable experimental scenario is described by (7.30) and corresponds to a doped structure with a finite Fermi level in the conduction band. For magnetic atoms placed in the vicinity of surfaces, terms proportional to *M* are negligible and the interaction is expressed as

$$\begin{aligned} H_{cc} &= w\big(Z_1, Z_2, R_{\|}\big) S_{1z} S_{2z}, \\ w\big(Z_1, Z_2, R_{\|}\big) &= \frac{L(Z_1, Z_2)}{4\pi^2\, n^2} T \sum_{\omega_n} I_{c\uparrow}(\omega_n) I_{c\downarrow}(\omega_n), \\ I_{c\uparrow\downarrow}(\omega_n) &= \int_0^\infty \frac{kdk\, J_0\big(kR_{\|}\big)}{i\omega_n - E_{c\uparrow\downarrow}(k) + \mu}. \end{aligned} \tag{7.33}$$

Detailed calculations in (7.33) are given in Appendix 2. The RKKY-type indirect exchange interaction via conduction electrons with well-defined Fermi surface takes the form:

$$w(Z_1, Z_2, R_{\|}) = -\frac{L(Z_1, Z_2)m}{4\pi n^2 \hbar^2 R_{\|}^2}\{I_1(R_{\|}) + I_2(R_{\|})\},$$

$$I_1 = \int_0^{R_{\|}k_R} s\Phi(s)P(s)ds, \quad I_2 = \int_{R_{\|}k_R}^{R_{\|}\sqrt{k_R^2+k_F^2}} s\Phi(s)Q(s)ds,$$

$$\Phi(s) = 1 - (R_{\|}k_R)^2/s^2,$$

$$P(s) = J_0(R_{\|}k_R + s)N_0(R_{\|}k_R - s) - J_0(R_{\|}k_R - s)N_0(R_{\|}k_R + s),$$

$$Q(s) = J_0(s - R_{\|}k_R)N_0(s + R_{\|}k_R) + J_0(s + R_{\|}k_R)N_0(s - R_{\|}k_R), \tag{7.34}$$

where $N_0(s)$ is the Newmann function (Bessel function of the second kind), $k_R = \alpha_R m/\hbar^2 = 2\tilde{A}_2 Vm/\hbar^2$ and $k_F = 2m\varepsilon_F/\hbar^2$ are the Rashba and the Fermi momentum, respectively.

7.5.1 Voltage-Controlled Magnetic Ordering

The Rashba momentum k_R (see Appendix 2) carries dependence on the vertical bias and makes the interaction (7.34) voltage-controlled. If the vertical bias is absent, the interaction is determined by the I_2-term in (7.34) and presents a standard oscillating 2D-RKKY range function [14, 15]. Under an applied voltage, the range function contains additional beatings induced by the Rashba spin-splitting. The oscillating nature of the I_2-term is illustrated in Fig. 7.9 in two instances: first, for the unphysical large Rashba momentum in order to visualize the beating as a mathematical structure of the I_2-term (Fig. 7.9a), and, second, for reasonably estimated spin-splitting (Fig. 7.9b).

The result is to be expected from the qualitative standpoint as the beating is a consequence of two distinct Fermi momenta in a Rashba spin-split electron gas. The integral I_1 is the contribution to the range function that relies exclusively on voltage-

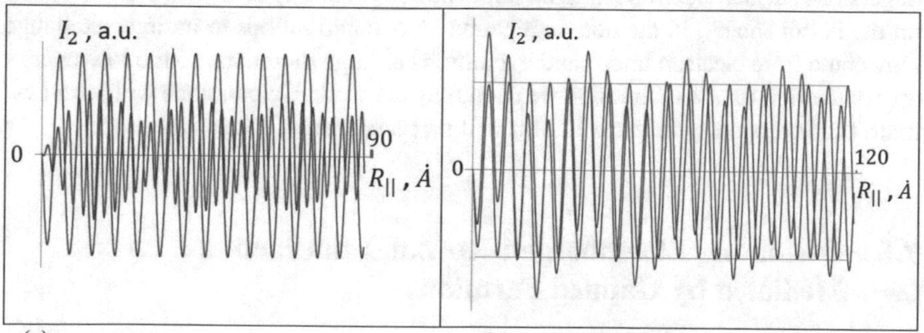

Fig. 7.9 Range function $I_2(R_{\|})$, $k_F = 0.5$ eV. Red line-$k_R = 0$. Blue line-$k_R \neq 0$. **(a)** $k_R = 1\ \dot{A}^{-1}$, **(b)** $k_R = 0.1\ \dot{A}^{-1}$

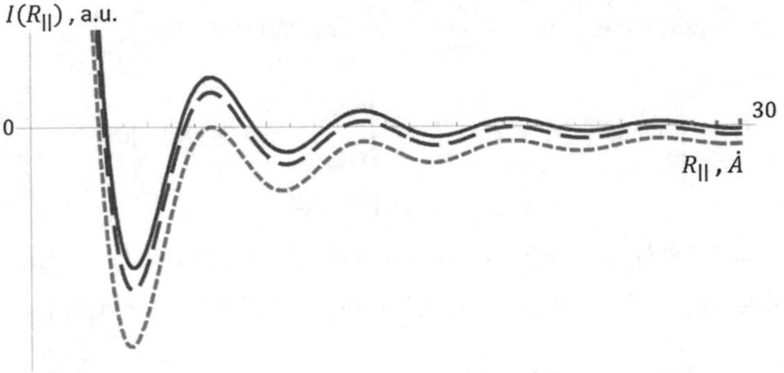

Fig. 7.10 Range function (7.34). Solid line-$k_R = 0$; Dashed line-$k_R = 0.02\ \dot{A}^{-1}$; Dotted line-$k_R = 0.05\ \dot{A}^{-1}$

induced spin-splitting. This term increases with applied voltage and, despite the fact that it experiences weak spatial oscillations, it makes the total interaction sign-definitive if the vertical voltage increases. This trend is illustrated in Fig. 7.10 where the total range function in (7.34), $I(R_{\parallel}) = -(I_1 + I_2)/R_{\parallel}^2$, is shown at different Rashba momenta.

An applied voltage increases the Rashba momentum and turns the sign-variable interaction into a sign-definitive one that favors ferromagnetic ordering at the surface. Coefficient $L(Z_1, Z_2)$ is responsible for the selection of the surface where the interaction takes place (see Figs. 7.6 and 7.7).

It should be noted that the spin-splitting-related features in RKKY may exist even without external bias as films are grown on a substrate and the built-in electric field makes surfaces electrically non-equivalent, thus violating the inversion symmetry.

If the chemical potential is placed within the surface energy gap, the inter-band terms (7.32) determine the magnetic ordering in the slab. As for microscopic calculations in (7.32), it should be noted that so far we have used an *s-d* model based on an approximate interaction matrix Q in the limit of $k \to 0$. It does not affect the result for degenerate surface electrons as main contributions to (7.34) come from the electrons on the Fermi sphere. In the non-degenerate case, contributions to indirect exchange may come from electron inter-band excitations of large momenta, so the low-energy approximation for *s-d* interaction we used may not work. Calculations with exact *s-d* matrix in inversion symmetric TI film will be performed in the next section.

7.6 Oscillating Bloembergen-Rowland Interaction Mediated by Gapped Fermions

If the Fermi level is far from the Dirac point, the oscillating RKKY interaction, calculated in the previous section, is not sensitive to whether surface electrons are topologically non-trivial or not. The energy gap, if accompanied by a loss of time reversal symmetry, turns a semiconductor to a trivial insulator state characterized by the zero-valued Z_2 topological invariant. However, opening a gap does not necessarily relate to a zero Z_2: the tunneling mixing of the states on opposite surfaces in a thin film may open an energy gap, preserving massive Dirac electrons in the topological state with $Z_2 = 1$ [16–18]. The coexistence of the gap and topological state affects the indirect exchange interaction which we discuss below.

As mentioned in the previous section, placing the chemical potential in the gap precludes using the small momentum approximation to *s-d* interaction, so an exact calculation of the indirect exchange should be done.

Below we start with the effective surface model (Chap. 6) in a representation that leaves the electron part of the Hamiltonian non-diagonal and the *s-d* interaction *k*-independent (see (6.8) and (6.10)). The free electron Hamiltonian projected onto surface states and its double spin degenerate eigenvalues have the form:

$$H_{\text{eff}}^{\text{el}} = \begin{pmatrix} \varepsilon_k & 0 & 0 & i\tilde{A}_2 k_- \\ 0 & -\varepsilon_k & i\tilde{A}_2 k_- & 0 \\ 0 & -i\tilde{A}_2 k_+ & \varepsilon_k & 0 \\ -i\tilde{A}_2 k_+ & 0 & 0 & -\varepsilon_k \end{pmatrix},$$

$$E_{\text{c,v}}(k) = \pm\sqrt{\varepsilon_k^2 + \tilde{A}_2^2 k^2}, \quad \varepsilon_k = \Delta_S/2 - Bk^2, \quad k_\pm = k_x \pm ik_y. \tag{7.35}$$

The condition $B\Delta_S > 0$ ensures that surface electrons are in a topologically non-trivial state. The *s-d* interaction is given as

$$H_{\text{eff}}^{\text{sd}} = \frac{1}{n_S} \sum_{ki, k'j} W_{ij}(Z) \exp\left(i(\boldsymbol{k} - \boldsymbol{k}')\boldsymbol{R}_\parallel\right) a_{ik}^+ a_{jk'},$$

$$W(Z) = \begin{pmatrix} J_{z1}S_z & J_{z2}S_z & 0 & 0 \\ J_{z2}S_z & J_{z1}S_z & 0 & 0 \\ 0 & 0 & -J_{z1}S_z & -J_{z2}S_z \\ 0 & 0 & -J_{z2}S_z & -J_{z1}S_z \end{pmatrix}, \tag{7.36}$$

where n_S is the sheet density of host atoms.

Compared to the Hamiltonian discussed in Chap. 6, here we deal with a structurally symmetric film $\left(\tilde{V}_{as} = 0\right)$ and neglect electron-hole asymmetry in the electron part of the Hamiltonian (7.35). Based on an analyses of the interaction constants in Chap. 6, we neglect the spin-flip *s-d* coupling J_t as it is proportional to the overlap of wave functions residing on opposite surfaces and reaches its maximum in the middle of the slab being an order of magnitude smaller than both intra- and inter-band spin-conserving constants J_{1z} and J_{z2}, respectively. If both local spins are placed in the vicinity of either surface, J_t plays no role.

The topological state is characterized by the spin degenerate inverted electron spectrum illustrated in Fig. 7.11.

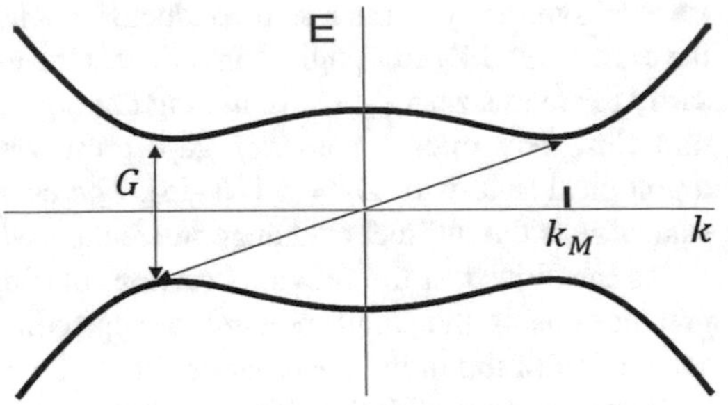

Fig. 7.11 Surface electron spectrum in a topological phase

The state is characterized by two parameters: the half-momentum that separates the two minima in the spectrum, k_{M}, and the minimum energy gap G at $k = k_{\mathrm{M}}$. Even without detailed calculations one can qualitatively predict the character of the range function of interacting impurity spins. As the BR indirect exchange interaction originates from the virtual electron-hole excitations across the energy gap [19], we have to account for the two types of excitations shown by arrows in Fig. 7.11: the direct transitions across the gap $G = \tilde{A}_2\sqrt{\left(2\Delta_S B - \tilde{A}_2^2\right)/B}$ as well as the indirect transitions with momentum transfer $2k_{\mathrm{M}}$, $k_{\mathrm{M}} = \sqrt{\left(\Delta_S B - \tilde{A}_2^2\right)}/B\sqrt{2}$. The direct transitions are responsible for the original BR-type interaction $\exp(-\alpha R)$, where the logarithmic decrement is determined by the bandgap G and the effective mass $m : \alpha = \sqrt{2mG}/\hbar$. The indirect transitions result in range oscillations on the scale of k_{M}^{-1}. So, we expect the modified BR interaction to be an oscillating range function with exponentially decreasing amplitude. As oscillations occur if k_{M} is a real number, $B\Delta_S > \tilde{A}_2^2$, one may conclude that the existence of the non-trivial topological state at $B\Delta_S > 0$, is the necessary condition for oscillations to occur.

Further below the above qualitative considerations will be confirmed by exact calculation of the BR indirect exchange, where the oscillation period and the decay scale will be expressed through the parameters of the surface electron spectrum. The inter-spin interaction is expressed as

$$
\begin{aligned}
H_{\mathrm{int}} = {} & \frac{T}{2(2\pi)^4 n_{\mathrm{S}}^2} \sum_{\omega_n} \iint dk\, dk' \\
& \times \exp\left[i\boldsymbol{R}_{\parallel}(\boldsymbol{k} - \boldsymbol{k}')\right] Tr\{W(Z_1)G(\boldsymbol{k}, \omega_n)W(Z_2)G(\boldsymbol{k}', \omega_n)\},
\end{aligned}
\tag{7.37}
$$

where $G(\boldsymbol{k}, \omega_n) = \left(i\omega_n - H_{\mathrm{eff}}^{\mathrm{el}}\right)^{-1}$.

We place the Fermi level at zero energy corresponding to the Dirac point at a single surface or to the middle of the tunneling gap in a thin film. Then the interaction takes the form

$$
\begin{aligned}
&H_{\text{int}} = w(\boldsymbol{R})S_{z1}S_{z2}, \quad w(\boldsymbol{R}) = \frac{T}{8\pi^4 n_{\text{S}}^2}(I_1 + I_2 + I_3),\\
&I_1 = C_1(Z_1, Z_2) \sum_{\omega_n} \int \Phi(\boldsymbol{k}, \boldsymbol{k}', \omega_n)\varepsilon_{\boldsymbol{k}}\varepsilon_{\boldsymbol{k}'} d\boldsymbol{k}\, d\boldsymbol{k}',\\
&I_2 = C_2(Z_1, Z_2) \sum_{\omega_n} \int \Phi(\boldsymbol{k}, \boldsymbol{k}', \omega_n)(i\omega_n)^2 d\boldsymbol{k}\, d\boldsymbol{k}',\\
&I_3 = -\tilde{A}_2^2 C_2(Z_1, Z_2) \sum_{\omega_n} \int \Phi(\boldsymbol{k}, \boldsymbol{k}', \omega_n) k k' \cos(\varphi - \varphi') d\boldsymbol{k}\, d\boldsymbol{k}',\\
&\Phi(\boldsymbol{k}, \boldsymbol{k}', \omega_n) = \frac{\exp\left[i\boldsymbol{R}_{\parallel}(\boldsymbol{k} - \boldsymbol{k}')\right]}{\left(\varepsilon_{\boldsymbol{k}}{}^2 + \tilde{A}_2^2 k^2 + \omega_n^2\right)\left(\varepsilon_{\boldsymbol{k}'}^2 + \tilde{A}_2^2 k'^2 + \omega_n^2\right)},\\
&C_{1,2}(Z_1, Z_2) \equiv J_{1z}(Z_1)J_{1z}(Z_2) \mp J_{z2}(Z_1)J_{z2}(Z_2), \quad \boldsymbol{k} = (k, \varphi).
\end{aligned}
\tag{7.38}
$$

The integration method in which the integration over momenta is followed by frequency summation was applied to the exchange interaction in narrow-gap semiconductors, metals, and graphene in [20–23]. A detailed example calculation of the integral I_2 is given in Appendix 3. Analytical calculations in (7.38) result in the indirect interaction expressed as

$$
\begin{aligned}
V(\boldsymbol{R}) = &\frac{G}{16\pi B^2 n_{\text{S}}^2}\{-C_1(Z_1, Z_2)F_1(R_{\parallel}) - C_2(Z_1, Z_2)F_2(R_{\parallel})\\
&+ \frac{2\,\tilde{A}_2^2\, C_2(Z_1, Z_2)}{BG} F_3(R_{\parallel})\},
\end{aligned}
\tag{7.39}
$$

where

$$
\begin{aligned}
&F_1(R_{\parallel}) = \int_1^{\infty} \frac{z\,dz}{\sqrt{z^2 - 1}} \left\{ \text{Im}\left[H_0^1\left(\frac{R_{\parallel} b(z)}{R_0}\right)\right]\right\}^2,\\
&F_2(R_{\parallel}) = \int_1^{\infty} \frac{\sqrt{z^2 - 1}}{z} \left\{ \text{Re}\left[H_0^1\left(\frac{R_{\parallel} b(z)}{R_0}\right)\right]\right\}^2 dz, \quad b(z) = \sqrt{R_0^2 k_{\text{M}}^2 + iz},\\
&F_3(R_{\parallel}) = \int_1^{\infty} \frac{dz}{z\sqrt{z^2 - 1}} \left\{ \text{Re}\left[b(z) H_1^1\left(\frac{R_{\parallel} b(z)}{R_0}\right)\right]\right\}^2, \quad R_0 \equiv \sqrt{2B/G}
\end{aligned}
\tag{7.40}
$$

and $H_n^1(z)$ are the Hankel functions of the first kind.

The interaction (7.39) depends on the positions of interacting spins along the z-axis through the coefficients $C_1(Z_1, Z_2)$ and $C_2(Z_1, Z_2)$ calculated for a four-quintuple layer film and shown in Fig. 7.12.

As illustrated in Fig. 7.12, for spins placed on the same surface, the coefficient $C_1(Z_1, Z_2)$ is negligibly small, being proportional to the top-bottom wave function overlap, so the indirect exchange interaction takes the form

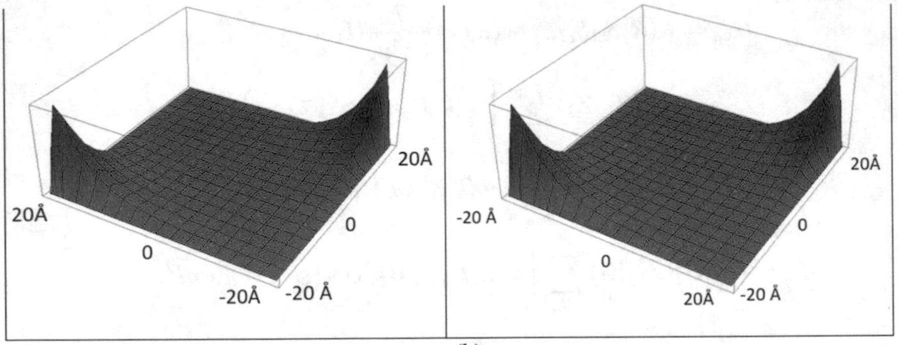

(a) (b)

Fig. 7.12 Spatial distribution of the exchange interaction along the direction normal to the plane: top ($Z_{1,2} = -20$ Å) to bottom ($Z_{1,2} = 20$ Å) axis. **(a)** $C_1(Z_1, Z_2)/J_z^2$, **(b)** $C_2(Z_1, Z_2)/J_z^2$

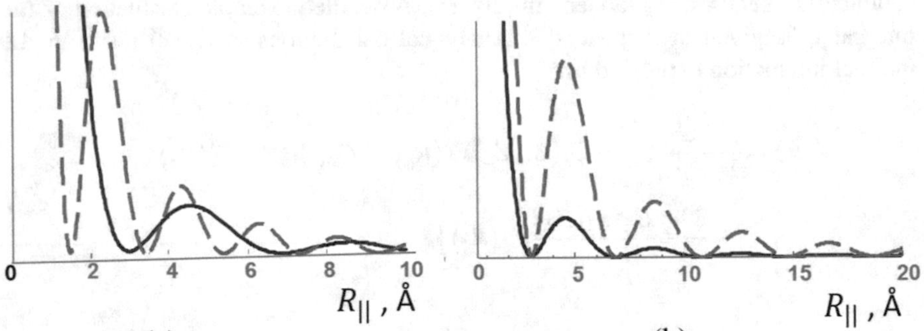

(a) (b)

Fig. 7.13 Function $F_2(R_{\|})$ (a. u.): **(a)** $R_0 = 5$ Å, solid line—$k_M = 0.8$ Å^{-1}, dashed line—$k_M = 1.6$ Å^{-1}; **(b)** $k_M = 0.8$ Å^{-1}, solid line—$R_0 = 5$ Å, dashed line—$R_0 = 10$ Å

$$V(\boldsymbol{R}) = -\frac{G\, C_2(Z_1, Z_2)}{16\pi B^2 n_S^2} \left\{ F_2\left(R_{\|}\right) - \frac{2\tilde{A}_2^2}{BG} F_3\left(R_{\|}\right) \right\}. \tag{7.41}$$

The functions $F_{1,2,3}(R_{\|})$ oscillate in the in-plane distance with exponentially decreasing amplitude. As an example, the function $F_2(R_{\|})$ is illustrated in Fig. 7.13. To clarify the character of range dependence, it is instructive to keep the parameters k_M and R_0 independent of each other and compare graphs for certain k_M and $2k_M$ (Fig. 7.13a) as well as for certain R_0 and $2R_0$ (Fig. 7.13b). This numerical comparison shows that a twofold increase in k_M leads to a twofold decrease in the distance between successive peaks, so the oscillation period is proportional to k_M^{-1}.

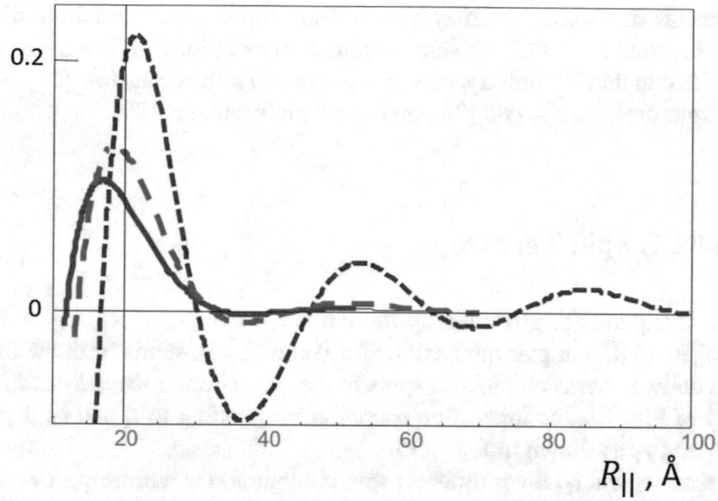

Fig. 7.14 Bloembergen-Rowland interaction oscillating with in-plane distance. Effective spin-orbit parameter $\tilde{A}_2\left[\text{eV}\dot{\text{A}}\right]$: solid line, 0.7; dashed line, 0.5; dotted line, 0.2

Similarly, a twofold increase in R_0 results in a twofold decrease in the rate of exponential decay, so R_0^{-1} determines the logarithmic decrement of spatial decay.

In fact, R_0 and k_M are expressed via basic parameters of the model and as such they are not independent. For the set of basic parameters, $\Delta_S = 0.2\ \text{eV}, B = 10\ \text{eV}\dot{\text{A}}^2$, the dimensionless figure bracket in (7.41) depends on the distance between magnetic atoms as shown in Fig. 7.14. A series of graphs is shown for various values of effective spin–orbit coupling that satisfy the constraint $B\Delta_S > \tilde{A}_2^2$.

The variable sign of interaction implies that competing ferro- and antiferromagnetic coupling between near-surface impurities may lead to the formation of various spin patterns. The film thickness affects the interaction through the tunneling gap Δ_S and thus controls the rate of exponential decay and the oscillation period: as compared to a thin film, the thicker one would have both a longer period and characteristic length of decay. These thickness-related trends may differentiate oscillations discussed here from those of the RKKY-type that might arise if the gap were filled with disorder-induced conductive states.

In conclusion, the inverted band spectrum of surface fermions in TI mediates the indirect exchange interaction that reveals long-ranged spatial oscillations even in a ground state with no real carriers at zero absolute temperature, so the Fermi surface and thus the RKKY exchange do not exist. Under these circumstances, the indirect exchange in a normal dielectric would follow the original Bloembergen-Rowland law—that is, monotonic exponential decay. The oscillating BR interaction is a signature of the topologically non-trivial state in a thin TI film. It should be noted that independent Dirac cones in thick films do not mediate oscillating spin–spin interaction. For a single surface, the interaction is of the RKKY-type in a metallic state, or is the one fading with distance either by power or exponential law when the chemical potential lies in the Dirac point or in the gap, respectively [24, 25].

The results discussed here may also concern topological crystalline dielectrics based on inverted band PbSnTe semiconductor alloys [26, 27]. The range function similar to that in thin TI films appears in indirect-gap semiconductors [11], graphene with the spin-orbit energy gap [28], and excitonic insulators [29].

Appendix 1. Spin Texture

The spin configuration, governed by the terms $t_1 = C_1(\boldsymbol{S_1} \times \boldsymbol{R}_{\|})_z(\boldsymbol{S_2} \times \boldsymbol{R}_{\|})_z$ and $t_2 = C_2(\boldsymbol{S_1}\boldsymbol{R}_{\|})(\boldsymbol{S_2}\boldsymbol{R}_{\|})$ in pair interaction (7.30) and (7.32), stems from the minimum energy it delivers when one rotates spins in the pair. If one rotates $\boldsymbol{S}_1$ and $\boldsymbol{S}_2$ in the geometry of Fig. 7.4, the interaction energy corresponding to t_1 and t_2, depends on angles γ_1 and γ_1 as shown in Fig. 7.15.

The terms, t_1 and t_2, favor different spin configurations with respect to the inter-spin vector (see insets in Fig. 7.15a, b) while they are similar in their antiferromagnetic nature as long as $C_{1,2} > 0$. If any of the coefficients $C_{1,2}$ becomes negative, the corresponding ordering turns ferromagnetic.

The actual spin texture in a magnetically doped TI film is determined by the total indirect exchange interaction that simultaneously includes competing t_1 and t_2 contributions as well as the Ising term in (7.30). In addition, the coefficients which are the range functions may be variable in sign. If one considers the magnetic ordering on a small inter-spin distance where sign variations are not relevant, the relative magnitude of the interaction on the surfaces and in the middle of the film make us conclude that surfaces experience out-of-plane ferromagnetism, while the spin glass forms in the inner part of the film.

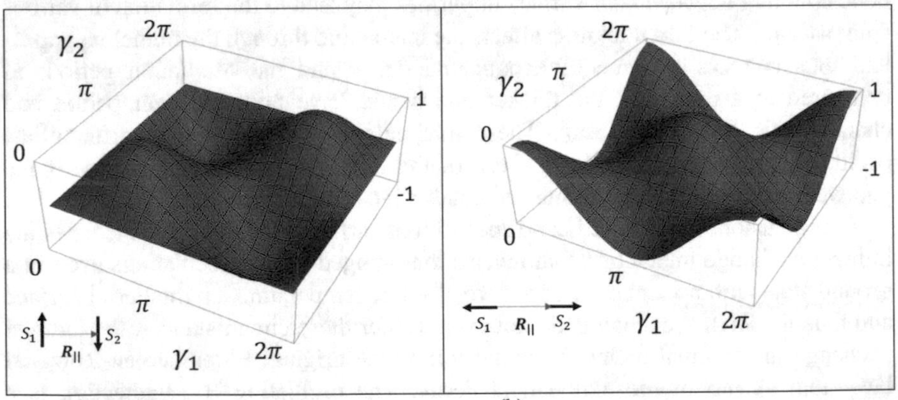

Fig. 7.15 t_1 **(a)** and t_2 **(b)** energy surfaces in (γ_1, γ_2) plane, $C_{1,2} > 0$. Insets show the spin alignment corresponding to the minimum energy

Appendix 2. Calculation of the RKKY in a Vertically Biased Topological Insulator

In order to calculate the range function (7.33) analytically, we use the conduction band energy spectrum (7.21) expanded up to the k^2 term (reference energy is in the middle of the surface energy gap):

$$E_{c,\uparrow\downarrow}(k) = \frac{1}{2}\sqrt{\Delta_S^2 + 4\tilde{V}_{as}^2} + \tilde{B}k^2 \pm \alpha_R k,$$

$$\tilde{B} = D - Bg_1(g_1^2 + 4V^2) + \frac{g_1^2 \tilde{A}_2^2}{\sqrt{\Delta_S^2 + 4\tilde{V}_{as}^2}} \equiv \frac{\hbar^2}{2m},$$

$$\alpha_R = 2\tilde{A}_2 V, \quad g_1 = \frac{\Delta_S}{\sqrt{\Delta_S^2 + 4\tilde{V}_{as}^2}}, \quad V = \frac{\tilde{V}_{as}}{\sqrt{\Delta_S^2 + 4\tilde{V}_{as}^2}}. \tag{7.42}$$

In what follows we suppose that $\tilde{B}$ (inverse effective mass) is positive and then the spectrum as a function of momentum comprises two parabolas shifted with respect to each other by Rashba spin-splitting ($\alpha_R \neq 0$).

We use an analytical continuation of Bessel functions to a negative k-half-plane,

$$J_0(kR_\parallel) = \frac{1}{2}\left(H_0^{(1)}(kR_\parallel) + H_0^{(2)}(kR_\parallel)\right),$$
$$H_0^{(1)}(ze^{i\pi}) = -H_0^{(2)}(z), \tag{7.43}$$

where $H_0^{(1,2)}(R_\parallel k)$ are Hankel functions [30]. The integral over momentum in (7.33) can be represented as an integral along the line shifted in the upper k-half-plane to avoid the branch cut of the function $H_0^{(1)}$ on the negative k-axis:

$$I_{c\uparrow}(\omega_n) = \frac{1}{2}\int_{-\infty+i\delta}^{\infty+i\delta} \frac{kH_0^{(1)}(kR_\parallel)dk}{i\omega_n - \tilde{B}k^2 - \alpha_R k + \varepsilon_F}, \quad \varepsilon_F = \mu - \frac{1}{2}\sqrt{\Delta_S^2 + 4\tilde{V}_{as}^2}. \tag{7.44}$$

Expression (7.44) accounts for the spin-up band only. To keep the integration over this band, we have to flip $\alpha_R \rightarrow -\alpha_R$ in the course of transition to a negative k half-axis. Because of Kramers degeneracy, mere reflection $k \rightarrow -k$ would correspond to transition to the spin-down band.

The Hankel function $H_0^{(1)}(z)$ is analytical in the upper k-half-plane and tends to zero on the arc $z \rightarrow \infty$. So, in order to calculate (7.44), we close the integration path in the upper half-plane. The poles of the integrand are

$$k_{1,2} = \frac{1}{2\tilde{B}}\left[-\alpha_R \pm \sqrt{\alpha_R^2 + 4\tilde{B}(i\omega_n + \varepsilon_F)}\right]. \tag{7.45}$$

Fig. 7.16 Poles of the integrand in (7.44) on the complex k-plane

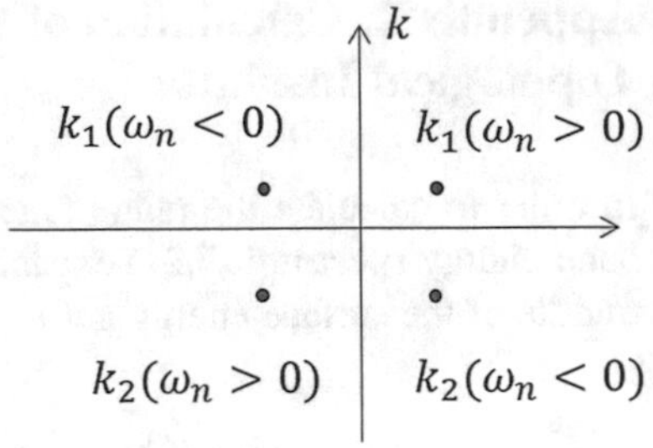

At this point, it is important to make sure that the poles of the integrand are located on both sides of the real k-axis. Otherwise, the integral would be zero as the path can be closed around the region, where the integrand is an analytical function. Moreover, as we will use (7.44) in a subsequent summation over the Matsubara frequency ω_n, it is important to choose branches of poles to hold their positions on the complex k-plane for $\omega_n > 0$ and $\omega_n < 0$, as shown in Fig. 7.16.

Finally, the integral (7.44) is found by calculating the residue

$$I_{c\uparrow}(\omega_n) = -\frac{\pi i}{2\tilde{B}} \cdot \frac{-P+\sqrt{P^2+ix+F}}{\sqrt{P^2+ix+F}} H_0^{(1)}\left[R_\parallel\left(-P+\sqrt{P^2+ix+F}\right)\right],$$
$$P = \frac{\alpha_R}{2\tilde{B}}; \quad F = \frac{\varepsilon_F}{2\tilde{B}}; \quad x = \frac{\omega_n}{2\tilde{B}}. \tag{7.46}$$

The integral $I_{c\downarrow}(\omega_n)$, calculated in a similar way, differs from (7.46) by the sign of coefficient P. The combination that enters the range function (7.33) takes the form:

$$I_{c\uparrow}(\omega_n)I_{c\downarrow}(\omega_n) = -\frac{\pi^2}{4\tilde{B}^2} \times \left(1 - \frac{P^2}{P^2+ix+F}\right) H_0^{(1)}\left[R_\parallel\left(-P+\sqrt{P^2+ix+F}\right)\right]$$
$$\times H_0^{(1)}\left[R_\parallel\left(P+\sqrt{P^2+ix+F}\right)\right], \tag{7.47}$$

The range function contains the sum over Matsubara frequencies which at $T \to 0$ can be replaced with the integral:

$$S_M = T\sum_{\omega_n} I_{c\uparrow}(\omega_n)I_{c\downarrow}(\omega_n) = \frac{1}{2\pi i}\int_\Gamma I_{c\uparrow}(\omega)I_{c\downarrow}(\omega)d\omega,$$
$$i\omega_n \to \omega + i\,sign(\omega). \tag{7.48}$$

Contour $\Gamma = \Gamma_1 + \Gamma_2$, shown in Fig. 7.17, avoids the branch cut point $x_0 = -R^2 - F$ and the branch cut line determined by the square root in (7.47).

The integral over the circle around the pole tends to zero when the radius decreases, and S_M can be expressed via integration over the upper and lower sides of path Γ_2:

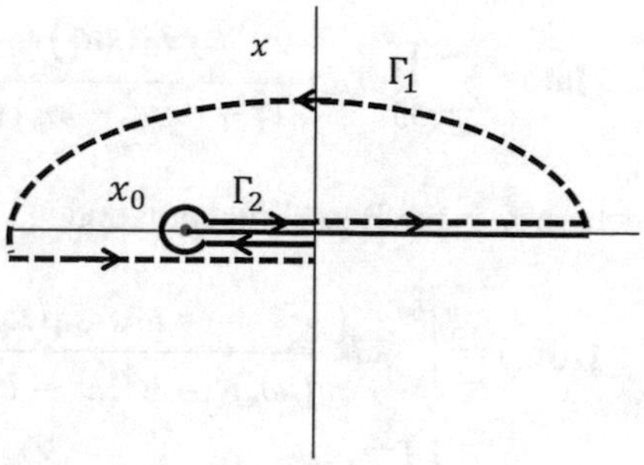

Fig. 7.17 Integration contour in (7.48)

$$S_{\mathrm{M}} = \frac{i\pi}{4\tilde{B}} \int_0^{\sqrt{P^2+F}} ydy\left(1 - \frac{P^2}{y^2}\right)\{H_0^{(1)}[R_\parallel(-P+y)]H_0^{(1)}[R_\parallel(P+y)]$$
$$- H_0^{(1)}[R_\parallel(-P-y)]H_0^{(1)}[R_\parallel(P-y)]\},$$
$$y = \sqrt{P^2 + x + F}. \tag{7.49}$$

The Hankel function $H_0^{(1)}$ has a branch cut on the negative part of the real axis. To avoid the integration of $H_0^{(1)}$ in this region, we use the relation $H_0^{(1)}(-y) = -H_0^{(2)}(y)$, so (7.49) takes the form

$$S_{\mathrm{M}} = \frac{i\pi}{4\tilde{B}} \times \left\{ \int_0^P yh_1(y)\left(1 - \frac{P^2}{y^2}\right)dy + \int_P^{\sqrt{P^2+F}} yh_2(y)\left(1 - \frac{P^2}{y^2}\right) \right\} dy,$$
$$h_1(y) = H_0^{(2)}[R_\parallel(P+y)]H_0^{(1)}[R_\parallel(P-y)] - H_0^{(2)}[R_\parallel(P-y)]H_0^{(1)}[R_\parallel(P+y)],$$
$$h_2(y) = H_0^{(1)}[R_\parallel(y-P)]H_0^{(1)}[R_\parallel(y+P)] - H_0^{(2)}[R_\parallel(y+P)]H_0^{(2)}[R_\parallel(y-P)]. \tag{7.50}$$

Changing the variable to $s = yR_\parallel$ and using notations (7.46), we get the final result for the range function (7.34) [31]. In (7.34), we used the identities relating to the Hankel and Newman functions: $H_0^{(1)}(s) = J_0(s) + iN_0(s), H_0^{(2)}(s) = J_0(s) - iN_0(s)$.

Appendix 3. Calculation of Bloembergen-Rowland Coupling

Below we illustrate the calculation method taking the integral I_2 (7.38) as an example. After integration over angles, we come to the expression

$$\mathrm{Int} = \sum_{\omega_n} \iint dkdq \frac{kq\, J_0\big(kR_\|\big) J_0\Big(qR_\|\Big) (i\omega_n)^2}{\big(\varepsilon_k^2 + \tilde{A}_2^2 k^2 - \omega_n^2\big)\big(\varepsilon_q^2 + \tilde{A}_2^2 q^2 - \omega_n^2\big)} \equiv \sum_{\omega_n} [Q(\omega_n)]^2, \quad (7.51)$$

where J_0 is the Bessel function and

$$\begin{aligned} Q(\omega_n) &= \int_0^\infty kdk\, \frac{i\omega_n\ J_0\big(R_\|\, k\big)}{(i\omega_n)^2 - B^2\big(k^2 - k_0{}^2\big)^2 - \tilde{A}_2^2 k^2} \\ &= \frac{1}{2}\int_0^\infty kdk\, \frac{i\omega_n}{(i\omega_n)^2 - B^2\big(k^2 - k_0{}^2\big)^2 - \tilde{A}_2^2 k^2} \Big[H_0^{(1)}\big(R_\|\, k\big) + H_0^{(2)}\big(R_\|\, k\big)\Big]. \end{aligned} \quad (7.52)$$

Using the relation $H_0^{(2)}\big(R_\|\, k\big) = -H_0^{(1)}\big(R_\|\, ke^{i\pi}\big)$, the second term in (7.52) can be represented as

$$\int_{-\infty+i\delta}^{0} ydy\, \frac{i\omega_n\, H_0^{(1)}\big(R_\|\, y\big)}{(i\omega_n)^2 - B^2\big(y^2 - k_0{}^2\big)^2 - \tilde{A}_2^2 y^2}. \quad (7.53)$$

Combining both terms in (7.52), we obtain

$$Q(x) = -\frac{1}{2B^2}\int_{-\infty+i\infty}^{\infty+i\infty} kdk\, \frac{x\, H_0^{(1)}\big(R_\|\, k\big)}{(k-k_1)(k-k_2)(k-k_3)(k-k_4)}, \quad (7.54)$$

where

$$k_{1,2} = \pm\left[k_{\mathrm{M}}^2 + \frac{1}{B}\sqrt{x^2 - \Delta_S^2/4}\right]^{1/2},\ k_{3,4} = \pm\left[k_{\mathrm{M}}^2 - \frac{1}{B}\sqrt{x^2 - \Delta_S^2/4}\right]^{1/2},\ x = i\omega_n. \quad (7.55)$$

Calculating the integral (7.54), one obtains

$$\begin{aligned} Q(x) &= \\ &-\frac{\pi i x\left[H_0^{(1)}\left(R_\|\left[k_{\mathrm{M}}^2 + \frac{1}{B}\sqrt{x^2 - \Delta_S^2/4}\right]^{1/2}\right) - H_0^{(1)}\left(-R_\|\left[k_{\mathrm{M}}^2 - \frac{1}{B}\sqrt{x^2 - \Delta_S^2/4}\right]^{1/2}\right)\right]}{4B\sqrt{x^2 - \Delta_S^2/4}} \\ &= -\frac{\pi i x\left[H_0^{(1)}\left(R_\|\left[k_{\mathrm{M}}^2 + \frac{1}{B}\sqrt{x^2 - \Delta_S^2/4}\right]^{1/2}\right) + H_0^{(2)}\left(R_\|\left[k_M^2 - \frac{1}{B}\sqrt{x^2 - \Delta_S^2/4}\right]^{1/2}\right)\right]}{4B\sqrt{x^2 - \Delta_S^2/4}}. \end{aligned} \quad (7.56)$$

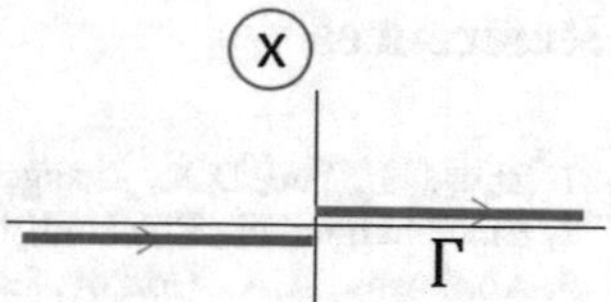

Fig. 7.18 Integration path on x-plane in (7.57)

At $T = 0$, discrete summation is replaced with integration over $x : i\omega_n \to x + i\,\delta\; sign\,(x)$, so the expression (7.56), being used in (7.51), gives

$$T\sum_{\omega_n}[Q(\omega_n)]^2=\frac{1}{2\pi i}\int_\Gamma [Q(x)]^2 dx. \qquad (7.57)$$

The integration path in (7.57) is shown in Fig. 7.18.

One can deform the contour Γ rotating it by $\pi/2$ as the turn goes in the region, where the integrand is an analytical function:

$$T\sum_{\omega_n}[Q(\omega_n)]^2=\frac{\pi i}{16B^2}\int_0^{i\infty}\frac{x^2}{x^2-\Delta_S^2/4}\left[H_0^{(1)}\left(R_\parallel\left[k_M^2+\frac{1}{B}\sqrt{x^2-\Delta_S^2/4}\right]^{1/2}\right)\right.$$
$$\left.+H_0^{(2)}\left(R_\parallel\left[k_M^2-\tfrac{1}{B}\sqrt{x^2-\Delta_S^2/4}\right]^{1/2}\right)\right]^2 dx. \qquad (7.58)$$

Changing the variable to $= \sqrt{1-4x^2/\Delta_S^2}$, we express (7.58) as

$$T\sum_{\omega_n}[Q(\omega_n)]^2=-\frac{\pi\Delta_S}{32B^2}\int_1^{\infty}\frac{\sqrt{z^2-1}}{z}\left[H_0^{(1)}\left(\frac{R_\parallel}{R_0}\left[R_0^2k_M^2+iz\right]^{1/2}\right)\right.$$
$$\left.+H_0^{(2)}\left(\frac{R_\parallel}{R_0}\left[R_0^2k_M^2-iz\right]^{1/2}\right)\right]^2 dz$$
$$=-\frac{\pi\Delta_S}{16B^2}\int_1^{\infty}\frac{\sqrt{z^2-1}}{z}\left\{\mathrm{Re}\left[H_0^{(1)}\left(\frac{R_\parallel}{R_0}\left[R_0^2k_M^2+iz\right]^{1/2}\right)\right]\right\}^2 dz. \qquad (7.59)$$

The expression (7.59) enters the final result (7.39). The terms I_1 and I_3 in (7.38) were calculated in a similar way.

References

1. Zhu, J.-J., Yao, D.X., Zhang, S.-C., Chang, K.: Phys. Rev. Lett. **106**, 097201 (2011)
2. Bloembergen, N., Rowland, T.J.: Phys. Rev. **97**, 1679 (1955)
3. Abrikosov, A.A., Gorkov, L.P., Dzyaloshinski, I.E.: Quantum Field Theoretical Methods in Statistical Mechanics. Pergamon, New York (1965)
4. Ruderman, M.A., Kittel, C.: Phys. Rev. **96**, 99 (1954)
5. Kasuya, T.: Prog. Theor. Phys. **16**, 45 (1956)
6. Yosida, K.: Phys. Rev. **106**, 893 (1957)
7. De Gennes, P.G.: J. Phys. Radium. **23**, 630 (1962)
8. Mattis, D.C.: The Theory of Magnetism Made Simple. World Scientific, Singapore (2006)
9. Bloembergen, N., Rowland, T.J.: Phys. Rev. **97**(6), 1680 (1955)
10. Litvinov, V.I.: Sov. Phys. Semicond. **19**, 345 (1985)
11. Abrikosov, A.A.: J. Low Temp. Phys. **39**(1/2), 217 (1980)
12. Liu, L., Bastard, G.: Phys. Rev. **B25**(1), 487 (1982)
13. Abanin, D.A., Pesin, D.A.: Phys. Rev. Lett. **106**, 136802 (2011)
14. Fischer, B., Klein, M.W.: Phys. Rev. **B11**, 2025–2029 (1975)
15. Litvinov, V.I., Dugaev, V.K.: Phys. Rev. **B58**(7), 3584 (1996)
16. Lu, H.-Z., Shan, W.-Y., Wang, Y., Niu, Q., Shen, S.-Q.: Phys. Rev. **115407**, B81 (2010)
17. Shan, W.-Y., Lu, H.-Z., Shen, S.-Q.: New J. Phys. **12**, 043048 (2010)
18. Shen, S.-Q., Shan, W.-Y., Lu, H.-Z.: arXiv:1009.5502v3 [cond-mat.mes-hall]
19. Litvinov, V.I., Dugaev, V.K.: Phys. Rev. Lett. **86**(24), 5593 (2001)
20. Dugaev, V.K., Litvinov, V.I.: Superlattices Microstruct. **16**, 413 (1994)
21. Litvinov, V.I., Dugaev, V.K.: Phys. Rev. B. **58**, 3584 (1998)
22. Dugaev, V.K., Litvinov, V.I., Barnas, J.: Phys. Rev. B. **74**, 224438 (2006)
23. Litvinov, V.: Wide Bandgap Semiconductor Spintronics. Pan Stanford, Singapore (2016)
24. Liu, Q., Liu, C.-X., Xu, C., Qi, X.-L., Zhang, S.-C.: Phys. Rev. Lett. **102**, 156603 (2009)
25. Efimkin, D.K., Galitski, V.: Phys. Rev. B. **89**, 115431 (2014)
26. Hsieh, T.H., Lin, H., Liu, J., Duan, W., Bansil, A., Fu, L.: Nat. Commun. **3**, 982 (2012)
27. Tanaka, Y., Ren, Z., Sato, T., Nakayama, K., Souma, S., Takahashi, T., Segawa, K., Ando, Y.: Nat. Phys. **8**, 800 (2012)
28. Dugaev, V.K., Litvinov, V.I., Barnas, J.: Phys. Rev. **B74**, 224438 (2006)
29. Litvinov, V.I.: Sov. Phys. Solid State. **27**(4), 740 (1985)
30. Abramovitz, M., Stegun, I.A. (eds.): Handbook of Mathematical Functions. National Bureau of Standards, Gaithersburg (1964)
31. Litvinov, V.I.: Phys. Rev. **B89**, 235316 (2014)

Chapter 8
Device Applications

Device applications of TI materials are based on their semiconductor properties and in this capacity they can be used in typical generic semiconductor electronic devices such as Schottky diodes, photodetectors, field-effect transistors, and thermoelectric units. In addition, the essentially topological spin-momentum locked states, which are absent in normal semiconductors, lead to the use of TI materials as a template for a variety of spintronic devices such as random access memory, information storage, and microwave sources.

Since the topological states are pinned to surfaces, the quality of contacts to various metals as well as the Fermi level position at the metal/TI interface affects the accessibility of the states, carrier mobility, and the quality of the Schottky barriers.

8.1 Contacts and Gating

Native defects (mostly chalcogen vacancies) in $Bi_2(Te_xSe_{1-x})_3$ create *n*-type doping and pin the Fermi level in the conduction band [1, 2]. The band bending near the surface forms an accumulation layer, thus creating a 2D electron gas and making it difficult to distinguish between 2D electrons and surface topological states. The density of the native defects depends on the growth method, post-growth treatment, as well as the choice of substrate and buffer layer. It was shown that proper post-growth annealing can reduce the density of chalcogen vacancies that move the Fermi level in the bulk bandgap close to the Dirac point [3].

The carrier density in an accumulation layer can be managed by electrostatic gating with metal contacts which could be either Ohmic or Schottky type, depending on the choice of metal. The active region of most optoelectronic devices consists of single or double heterojunctions characterized by variable energy gaps, high carrier

V. Litvinov, *Magnetism in Topological Insulators*,
https://doi.org/10.1007/978-3-030-12053-5_8

mobility, and the ability to confine nonequilibrium carriers. From this perspective, the various heterojunctions that a TI could form with the normal semiconductors are under intensive study. Several examples of contacts and heterojunctions are given below.

First-principle calculations, performed for Bi_2Se_3/(Au, Pt, Ni, Pd, graphene) contacts predict n-type Ohmic contacts only, regardless of the choice of metals from the group. Graphene and Au were found to have the weakest charge transfer to TI, thus providing the best Ohmic contacts which do not interfere with the topological states and preserve their spin-momentum locked character [4]. To prevent charge transfer to other metals, it was proposed to use a thin large bandgap dielectric layer placed between a metal and a TI. This type of contact is normally used as a gate to achieve surface conduction that is tunable from n- to p-type by gate voltage. Effective gating has been reported for $(Bi_{1-x}Sb_x)_2Te_3$ in a geometry with top [5, 6] and back (through the dielectric substrate) gates [7]. To get ambipolar surface conduction, the bulk conductivity should be suppressed as much as possible by the choice of alloy composition, i.e., by the Bi/Sb or (and) Se/Te ratio. The compositional tuning of the Fermi level makes the topological states accessible at $0.65 \leq x \leq 0.73$ in $(Bi_{1-x}Sb_x)_2Se_3$ [8] and $y = 0.94$ in $(Bi_{1-y}Sb_y)_2Te_3$ [9]. The p- to n-type conductivity transition around $y = 0.94$ is at the origin of the topological p-n junction proposal [10]. Manipulations of the Fermi level by alloy composition with subsequent observation of gate-controlled ambipolar conductivity were reported in $(Bi_{1-x}Sb_x)_2Te_3$ [11] and $(Bi_{1-x}Sb_x)_2Te_{1.25}Se_{1.75}$ [12].

8.2 Heterojunctions

The device quality of a semiconductor heterojunction relies on the structural quality of the interface and on the electrical parameters, such as band offsets, Schottky barrier height, and electron (hole) mobility. In the heterojunction TI/(normal semiconductor), the values of band offsets indicate whether the topological states are accessible, thus helping to engineer the performance of the device.

The heterojunctions consisting of a TI and semiconductors of group-IV and III–V Si, Ge, (Al, Ga, In)As, (Al, In, Ga)N are of special interest as they help integrate TIs into well-developed semiconductor templates and expand the functionalities of optoelectronic devices and field-effect transistors. Several examples of such heterojunctions are given below.

Group IV semiconductors. The band diagram of the interface Si(111)/Bi_2Se_3 is illustrated in Fig. 8.1. In the diagram, we use a barrier height of $\phi_B = 0.31$ eV, valence band offset of $\Delta E_v = 0.19$ eV, and Fermi energy, $E_F = 0.37$ eV [13].

For the surface orientation Si(001), a higher Schottky barrier of $\phi_B = 0.34$ eV was reported in [14]. The discrepancy between the values of ϕ_B found in [13, 14] may be attributed to the different Si-face orientations used in the experiments, and also to the hydrogen-passivation of the Si(111) surface, that was aimed at improving interface

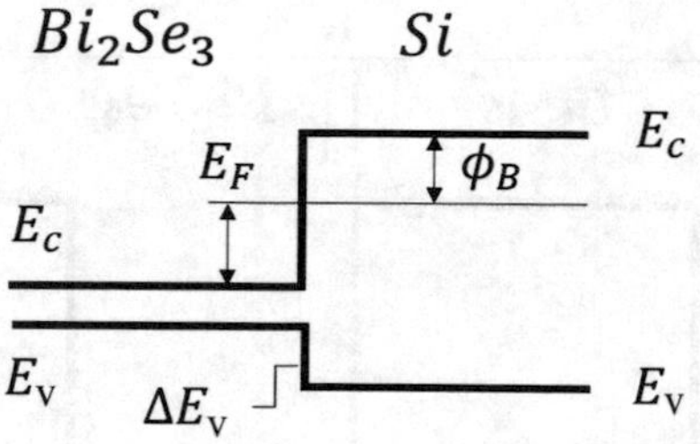

Fig. 8.1 Barrier Schottky at n - Si(111)/n - Bi_2Se_3 interface [13]

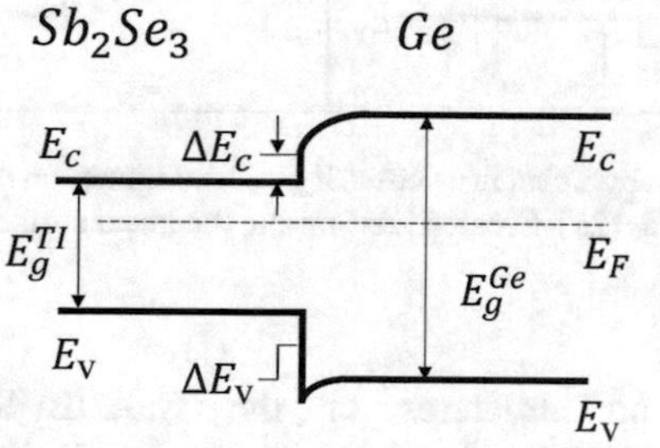

Fig. 8.2 Band alignment in Ge/Sb_2Te_3 heterojunction [16]. $E_g^{TI} = 0.35$ eV,$E_g^{Ge} = 0.67$ eV

quality and exclusion of defects in the formation of the Schottky barrier. As the Fermi level in both examples is pinned deeply in the TI conduction band, the topological states would hardly be accessible, so they do not affect the performance of Schottky devices. The Schottky barrier height depends on where the Fermi level is pinned at the interface, and for n-Si(111)/p-Bi_2Se_3 heterojunctions used in photodetector devices, the barrier height was estimated as 0.65 eV [15].

In a germanium contact to Sb_2Te_3, the conduction and valence band offsets were estimated as $\Delta E_c = 0.07$ eV and $\Delta E_v = 0.25$ eV, respectively. The band diagram is shown in Fig. 8.2.

The Fermi level position in the TI bandgap allows topological states to take part in the in-plane transport. The spin-locked character of topological states is preserved as the Ge/Sb_2Te_3 interface causes weak disturbance to the states [16].

Cubic III–V semiconductors. The GaAs substrates in the (001) [17, 18] and (111)B [19, 20] surface orientations are often used for TI epitaxial growth. GaAs is not the best substrate as it has a lattice mismatch as large as 3.4%, 5.9%, and 8.7% for Bi_2Se_3, Sb_2Te_3, and Bi_2Te_3, respectively [21]. Lattice mismatch causes compressive strain in coherently grown layers and may result in generating defects if the layer thickness exceeds the critical value. Despite this, the attractive side of using a GaAs substrate is that it favors the integration of a TI into a template suitable for high-speed electronic and optoelectronic devices.

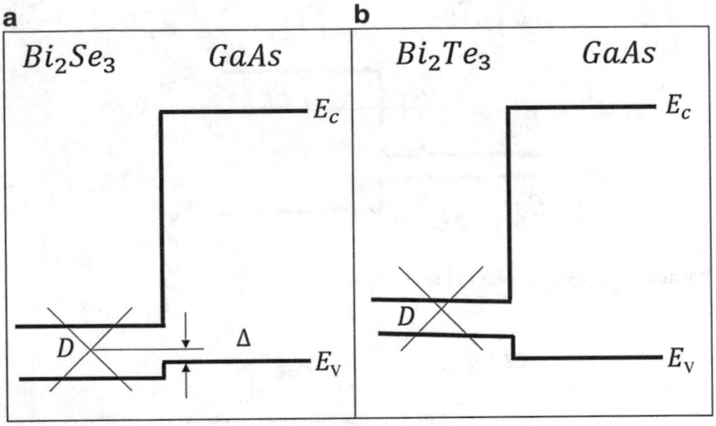

Fig. 8.3 (**a**) Type-II band alignment in a GaAs/Bi_2Se_3 heterojunction $\Delta = 0.05$ eV [22]. (**b**) Type-I heterojunction GaAs/Bi_2Te_3 [23]. Crossed lines mimic the linear surface spectrum in a TI. E_D is the Dirac point level

The interface band structures of the GaAs/Bi_2Se_3 and GaAs/Bi_2Te_3 heterojunctions are shown in Fig. 8.3a, b. The band alignment follows from a simplified consideration of conduction and valence level positions relative to vacuum. The type-II heterojunction shown in Fig. 8.3a was studied by first-principle calculations in [22]. It was shown that the hybridization between the topological and GaAs valence states at the interface adds momentum-dependent spin texture to non-topological states.

Circularly polarized light with interband photon energy $\hbar\omega \geq E_g$(GaAs) excites spin-polarized electrons on both sides of the junction shown in Fig. 8.3b. If the TI film is thin enough, a considerable part of the optical power penetrates Bi_2Te_3 and is absorbed in the GaAs substrate. After the spin-polarized hot electrons are injected across the heterojunction into Bi_2Te_3, they create a spin imbalance. This kind of spin injection has proven to contribute to the photocurrent due to the circular photogalvanic effect [23, 24]. The photo-excited topological insulator heterostructure GaAs/$(Bi_{0.5}Sb_{0.5})_2Te_3$ has been considered as a possible broadband terahertz source which converts very fast transient spin current to charge current due to the inverse spin Hall effect [25].

II–VI Semiconductors. In type-I heterointerface Bi_2Te_3/CdTe(111)B, the band diagram is similar to that rendered in Fig. 8.1, where the conduction and valence band offsets, $\Delta E_c = 1.12$ eV and $\Delta E_v = 1.12$ eV, respectively, follow from the spectra of photoelectron spectroscopy in the X-ray and ultraviolet spectral regions [26].

Wurtzite III–V semiconductors. Wide bandgap III–V semiconductors are widely used in high-power field-effect transistors and various optoelectronic devices such as light-emitting diodes and lasers. The combination of two groups of materials,

AlGaN-based semiconductors and bismuth chalcogenides, may open new possibilities in device applications, expanding their area of functionality. The AlN/Si(111) wafers are well-developed industry-grade substrates which can be used to integrate the epitaxial TI layers into existing Si-based electronic devices. In this context, important steps have been taken to grow the thin epitaxial Bi_2Se_3 layers directly on the AlN/Si(111) substrate, resulting in layer quality high enough to preserve the gapless topological surface states [27].

The electrical properties of the interface between AlN and TI are characterized by large conduction and valence band offsets, as illustrated in Fig. 8.4. The band offsets were estimated using the conduction band affinity to vacuum with no regard to the charge transfer across the junction.

In heterojunctions with $Al_{1-x}Ga_xN$, the band diagram is tuneable with composition x. In the Bi_2Te_3/GaN heterointerface the conduction band offset is negative, manifesting type-II (staggered) alignment, as shown in Fig. 8.5.

The GaN substrate by itself presents a template for various visible wavelength optoelectronic devices. Had a TI heterojunction been grown on the same substrate, it would have been an integrated infrared (IR) photodetector with a Bi_2Te_3 active region. IR photodetectors based on Bi_2Te_3 are discussed in Sect. 8.3.

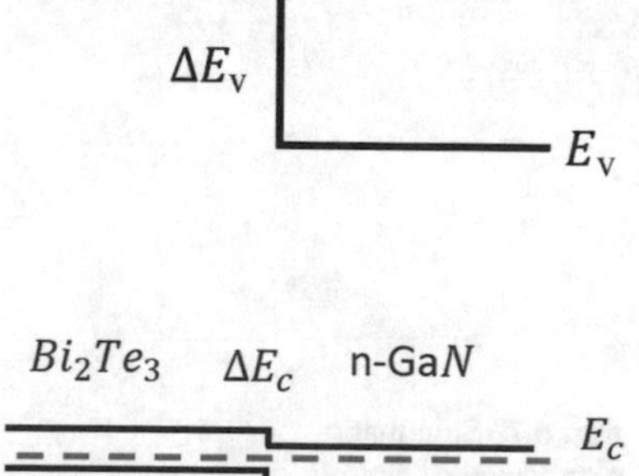

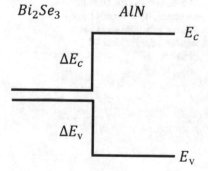

Fig. 8.4 Bi_2Se_3/AlN heterojunction. $\Delta E_c = 2.86$ eV, $\Delta E_v = 2.87$ eV [28]

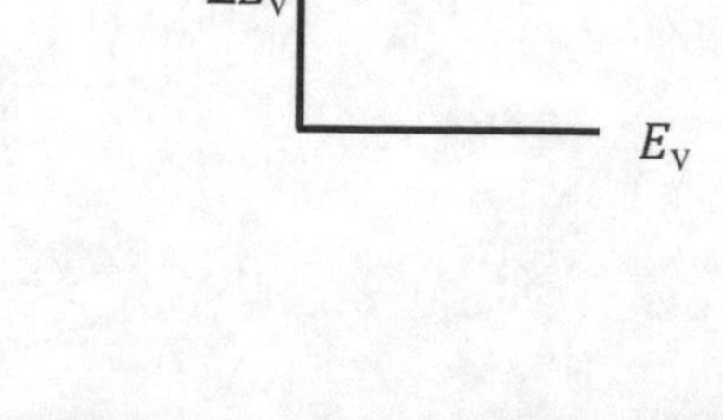

Fig. 8.5 Type-II heterojunction Bi_2Te_3/GaN. $\Delta E_c = -0.05$ eV, $\Delta E_v = 3.27$ eV [29]

8.3 Photodetectors

Any semiconductor can be used as a photodetector for the wavelength corresponding to the bandgap: $\lambda(\mu m) < 1.24/E_g(eV)$. Topological insulators are gapped in the bulk and are gapless or narrow-gapped at the surface, depending on the thickness. The small gap at the surface in $Bi_2(Te, Se)_3$ is at the origin of the broadband optical absorption and photosensitivity over the infrared to terahertz spectral range.

Various types of photodetectors with TI active regions have been fabricated on various substrates. Examples are given below.

The *p-n* heterostructure Bi_2Te_3/GaN is shown in Fig. 8.5. The photoresponse upon illumination with a standard source ($\lambda = 1$ μm) has been studied in [30]. The quality of the *p-n* junction and the IR responsivity were recognized as promising in the device illustrated in Fig. 8.6.

In a normal semiconductor with a bandgap of 0.21 eV (the bandgap of bulk Bi_2Te_3) maximum performance would be expected in the range $\lambda < 5.9$ μm, where the inter-band transitions come into effect. Gapless surface electrons in TIs extend the sensitivity of the device into the far-infrared region, and as far as light is absorbed by transitions in the gapless linear spectrum, the absorbance does not depend on wavelength [31].

A generic photoconductive detector is shown in Fig. 8.7.

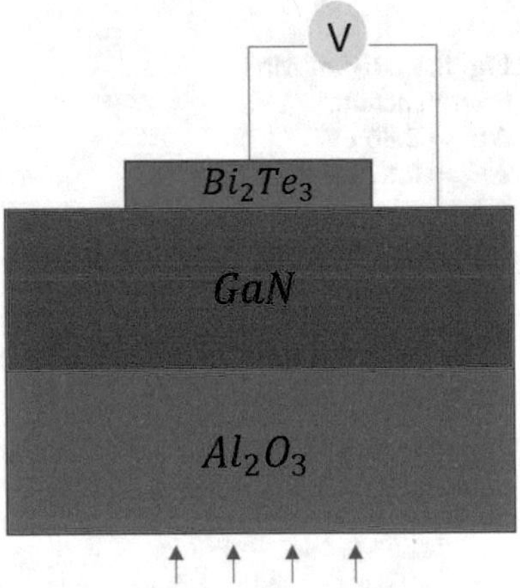

Fig. 8.6 Back-side illuminated Bi_2Te_3/GaN IR detector [30]. The GaN layer and the substrate serve as windows for incident IR radiation

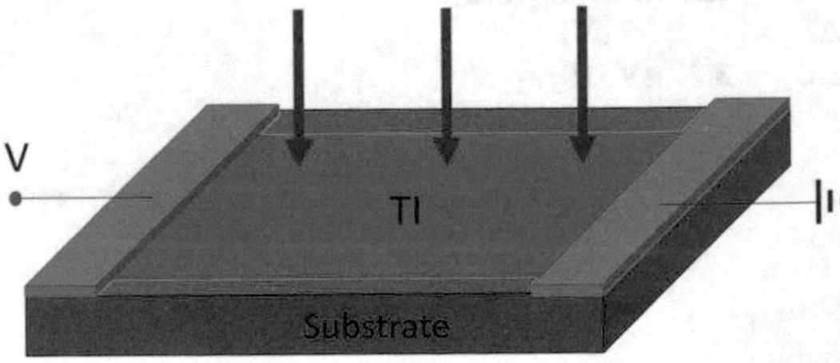

Fig. 8.7 Schematic photodetector setting

Molecular beam epitaxial growth of a Sb_2Te_3 photoconductive detector on an Al_2O_3 substrate shows the maximum responsivity and specific detectivity of $R = 21.7$ A/W and $D^* = 1.22 * 10^{11}$ cm$\sqrt{Hz}$/W, respectively, at $\lambda = 980$ nm at a temperature of 150 K [32]. In a device with an active region made of Sb_2SeTe_2 nanoflakes, room temperature responsivity was measured as high as 2293 A/W at $\lambda = 532$ nm [33].

Si-based TI detectors attract considerable attention as they can be integrated with other electronic components using industry-grade templates of high structural quality. High responsivity of 300 A/W at room temperature and fast operation have been reported for a photoconductor based on Bi_2Se_3 nanowire deposited onto a Si/SiO_2 substrate. Improved characteristics were attributed to strong electron quantum confinement in the wire [34].

The Bi_2Te_3/Si photoconductive device demonstrated room temperature responsivity of 3.64 mA/W at $\lambda = 1.064$ μm [35]. The *p-n* junction in the heterostructure (*p*)Bi_2Te_3/(*n*)Si spatially separates the non-equilibrium carriers, preventing fast recombination and increasing their lifetime. The broadband room temperature responsivity of the vertical *p-n* Bi_2Te_3/Si device has been experimentally observed in the spectral range from ultraviolet to terahertz, at $\lambda = (0.37, 1.55, 118.8)$μm and maximum responsivity and specific detectivity of 1 A/W and $2.5 * 10^{11}$ cm$\sqrt{Hz}$/W, respectively [36]. So far, the best performance for a heterojunction device has been reported in Bi_2Se_3/Si [15]: a responsivity of 24.28 A/W and detectivity of 4.39×10^{12} cm$\sqrt{Hz}$/W in the range from ultraviolet to the optical communication band $(0.35 - 1.1)$ μm.

There are photodetector devices in which the topological insulators are used in combination with graphene [37] and WSe_2 [38]. Parameters of various TI-based photodetectors have been compared in the review article [39].

The insulating bulk and the conductive surface make TI a unique material for guiding light. The simultaneous presence of the high refractive index in the bulk and Dirac plasmon excitations on the surface allows the realization of new nanostructures which increase light absorption in solar cells and photodetectors. In this respect, the surface of $Bi_{1.5}Sb_{0.5}Te_{1.8}Se_{1.2}$ patterned in nano-cones has improved the efficiency of solar cells due to the plasmonic resonance in the visible spectral range. It has proven to serve as an antireflection coating [40].

8.4 Field-Effect Transistors

Similarly to photodetectors, field-effect transistors (FETs) can be made of virtually any semiconductor. The use of TI as a channel material is driven by the hope to maintain high electron mobility due to suppressed backscattering of the surface modes. An increase in mobility was demonstrated when the electrostatic gating pushed the Fermi level close to the Dirac point, thereby increasing the contribution of topological states in the total carrier mobility [41]. Generic FET geometry is illustrated in Fig. 8.8.

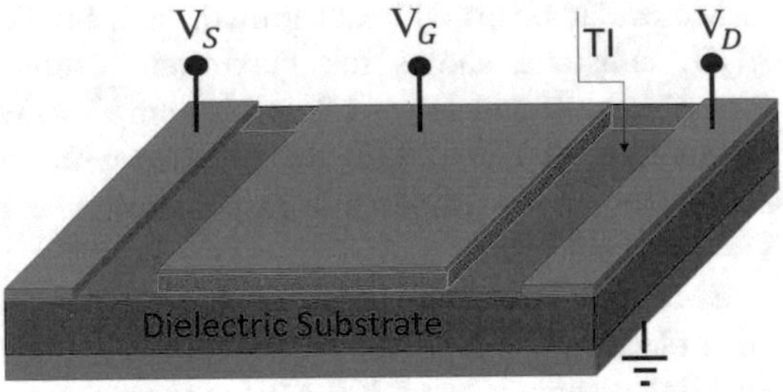

Fig. 8.8 Metal-oxide-semiconductor field-effect transistor. Gate dielectric is shown in yellow

As mentioned in Sect. 8.1, gating affects the conductivity in the channel and, basically, this is enough for the structure to operate as a transistor. However, if the channel is a TI, the absence of energy gap on the upper surface allows the source-drain current to flow, even if the gate bias is lower than the threshold value. This degrades the pinch-off characteristics of an FET. For this particular reason, TI layers normally used in FETs are thin enough to create an energy gap due to electron tunneling between the top and the bottom surfaces. An ultrathin FET with a clear OFF state at negative gate voltage has been demonstrated in a $Bi_2Se_3/SiO_2/Si$ structure [42]. The performance of the Bi_2Se_3 nanowire device shows the sharp pinch off which is attributed to the destruction of the gapless conduction channel by negative gate voltage [43].

To achieve high-frequency performance of TI-based FETs, the issue to be addressed relates to the fact that upper and lower surfaces are coupled by TI-layer capacitance. Despite manipulating the gate voltage, one can tune the upper surface to an almost depleted state, the bottom surface remains conductive as the vertical electric field makes the surfaces non-equivalent [44]. As Bi_2Se_3 has a high refractive index, the high capacitance of a TI layer may shunt the top and upper surfaces. In order to tune the top and bottom surfaces independently to a dielectric state, a scheme with two (top and bottom) gates is being used [45, 46].

8.5 Magnetic Devices

Current-induced magnetization switching is a key effect that underlies the operation of magnetoresistive random access memory and memory-in-logic integrated circuits. The main blocks of these devices are magnetic tunnel junctions (MTJs) which are comprised of two ferromagnetic metal layers, one with fixed and the other with free magnetization, separated by a thin non-magnetic dielectric. The spin-transfer torque (STT) in MTJ causes magnetization switching at a vertical current density on the order of 10^3 kA/cm^2.

Another type of torque is related to spin-orbit interaction and appears when the current flows parallel to an interface. That is why spin-orbit torque (SOT) requires a different device geometry compared to vertical STT devices. Possible options for STT- and SOT-based switch geometry and their modes of operation can be found in [47–52].

It was found that TI-based heterostructures demonstrate large SOT and allow switching at lower currents, so TI-based devices could replace MTJs and other non-TI SOT- and STT-based units [53]. The ferromagnetic material in a bilayer TI/FM can also be made of a magnetically doped TI. As an example, the bilayer $(Bi_{0.5}Sb_{0.5})_2Te_3/(Cr_{0.08}Bi_{0.54}Sb_{0.38})_2Te_3$ reveals switching current reduced down to 89 kA/cm^2 at low temperature [54]. That is much lower than the current in heavy metal/FM bilayers. Large SOT was also observed in a Bi_2Se_3/NiFe heterostructure [55]. The Bi_2Se_3/NiFe switch operates at room temperature providing a threshold current density as low as $6 * 10^2$ kA/cm^2 [56]. Both room temperature operation and current density at least one order of magnitude lower than that in non-TI devices promise that TI-based switches will become an attractive option for logic units and memory cells because of their low power consumption.

In an MTJ, besides magnetization switching in a free magnetic layer, the steady-state precession of the magnetic moment takes place at a particular set of geometry and material parameters. It happens when magnetic damping is compensated by an STT. The precession generates AC current in the GHz frequency range through a giant magnetoresistance effect [57]. Such an MTJ is a spin-transfer oscillator (STO) which can be used as a microwave source. To increase the AC output power, it is required that multiple STOs be electrically connected in series and synchronized. There is a domain of parameters where an AC signal produced by every single STO creates the feedback that leads to phase locking in the whole array. The synchronization does not require direct interaction between magnetic moments in neighboring STO. The process was studied for an array of two [58, 59] and of N [60–64] STO devices, scaling the output AC power as N^2.

In a TI-STO, the anomalous Hall effect and spin-momentum locking play major roles in the nonlinear magnetic dynamics of the device. Self-consistent theoretical analysis predicts periodic and aperiodic oscillations depending on the input parameters [65–67].

Low power consumption in a TI-STO promises to facilitate highly energy efficient arrays of phase-locked oscillators which can be used in phase-logic computing [68, 69]. The theoretical analysis of the performance of the coupled oscillator geometry shown in Fig. 8.9 was carried out in [70].

An individual STO presents a nonlinear system in which current-induced spin polarization affects FM magnetization through exchange coupling and torque. In turn, the precessing FM magnetization causes an anomalous Hall effect in a TI and acts directly on electron spin polarization, thus affecting the TI current, and

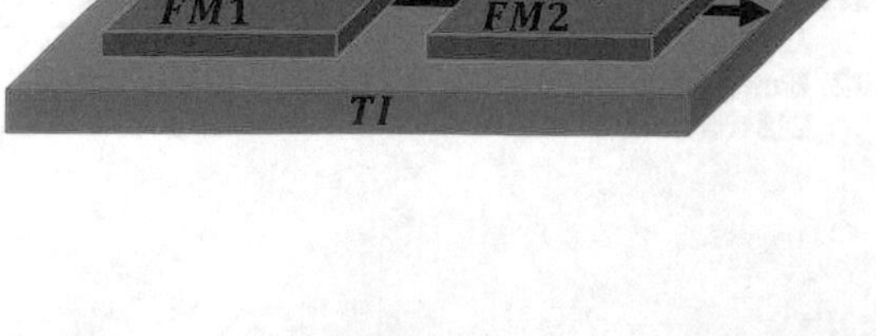

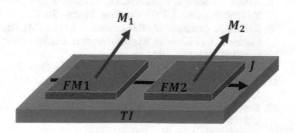

Fig. 8.9 Phase-locked precession of magnetic moments in a TI-STO array

maintaining the feedback from an FM. In a particular range of system parameters, the moment experiences steady-state precession. Moments ***M1*** and ***M2*** taken separately oscillate at different frequencies. However, a common current provides nonlinear coupling between them and synchronizes the oscillators. The operation of a phase-locked array is expected for a TI-STO made of the material combinations that follow: Bi_2Se_3/YIG($Y_3Fe_5O_{12}$), Bi_2Te_3/GdN, Bi_2Se_3/EuS, and Bi_2Se_3/$Cr_2Ge_2Te_4$.

8.6 Optoelectronics

As in the optical frequency range, the refractive indexes in the bulk and on the surface are different, a TI film can be utilized as a waveguide or an optical cavity [71].

The presence of almost gapless branches at the Bi_2Se_3 surface increases the number of absorption bands in the visible and mid-infrared spectral ranges [72]. Saturable absorption makes TI a material of choice for ultrafast Q-switched lasers in communication and mid-infrared bands. A detailed review of TI applications in lasers can be found in [39, 73].

References

1. ViolBarbosa, C.E., Shekhar, C., Yan, B., Ouardi, S., Ikenaga, E., Fecher, G.H., Felser, C.: Phys. Rev. **B88**, 195128 (2013)
2. Suh, J., Fu, D., Liu, X., Furdyna, J.K., Yu, K.M., Walukiewicz, W., Wu, J.: Phys. Rev. **B89**, 115307 (2014)
3. Walsh, L.A., Green, A.J., Addou, R., Nolting, W., Cormier, C.R., Barton, A.T., Mowll, T.R., Yue, R., Lu, N., Kim, J., Kim, M.J., LaBella, V.P., Ventrice Jr., C.A., McDonnell, S., Vandenberghe, W.G., Wallace, R.M., Diebold, A., Hinkle, C.L.: ACS Nano. **12**, 6310 (2018)
4. Spataru, C.D., Leonard, F.: Phys. Rev. **B90**, 085115 (2014)
5. Steinberg J.-B. Laloe, H., Fatemi, V., Moodera, J.S., Jarillo-Herrero, P.: Phys. Rev. **B84**, 233101 (2011)
6. Yang, F., Taskin, A.A., Sasaki, S., Segawa, K., Ohno, Y., Matsumoto, K., Ando, Y.: Appl. Phys. Lett. **104**, 161614 (2014)
7. He, X., Guan, T., Wang, X., Feng, B., Cheng, P., Chen, L., Li, Y., Wu, K.: Appl. Phys. Lett. **101**, 123111 (2012)
8. Satake, Y., Shiogai, J., Takane, D., Yamada, K., Fujiwara, K., Souma, S., Sato, T., Takahashi, T., Tsukazaki, A.: J. Phys. Condens. Matter. **30**, 085501 (2018)
9. Kellner, J., Eschbach, M., Kampmeier, J., Lanius, M., Młynczak, E., Mussler, G., Holländer, B., Plucinski, L., Liebmann, M., Grützmacher, D., Schneider, C.M., Morgenstern, M.: Appl. Phys. Lett. **107**, 251603 (2015)
10. Wang, J., Chen, X., Zhu, B.-F., Zhang, S.-C.: Phys. Rev. **B85**, 235131 (2012)
11. Kong, D., Chen, Y., Cha, J.J., Zhang, Q., Analytis, J.G., Lai, K., Liu, Z., Hong, S.S., Koski, K. J., Mo, S.-K., Hussain, Z., Fisher, I.R., Shen, Z.-X., Cui, Y.: Nat Nanotechnol. **6**, 705 (2011)
12. Banerjee, A., Sundaresh, A., Majhi, K., Ganesan, R., Kumar, P.S.A.: Appl. Phys. Lett. **109**, 232408 (2016)

13. Li, H., Gao, L., Li, H., Wang, G., Wu, J., Zhou, Z., Wang, Z.: Appl. Phys. Lett. **102**, 074106 (2013)
14. Ojeda-Aristizabal, C., Fuhrer, M.S., Butch, N.P., Paglione, J., Appelbaum, I.: Appl. Phys. Lett. **101**, 023102 (2012)
15. Zhang, H., Zhang, X., Liu, C., Lee, S.T., Jie, J.: ACS Nano. **10**, 5113 (2016)
16. Zheng, B., Sun, Y., Wu, J., Han, M., Wu, X., Huang, K., Feng, S.: J. Phys. D Appl. Phys. **50**, 105303 (2017)
17. Liu, X., Smith, D.J., Fan, J., Zhang, Y.-H., Cao, H., Chen, Y.P., Leiner, J., Kirby, B.J., Dobrowolska, M., Furdyna, J.K.: Appl. Phys. Lett. **99**, 171903 (2011)
18. Liu, X., Smith, D.J., Cao, H., Chen, Y.P., Fan, J., Zhang, Y.-H., Pimpinella, R.E., Dobrowolska, M., Furdyna, J.K.: J. Vasc. Sci. Technol. **B30**, 02B103 (2012)
19. Richardella, A., Zhang, D.M., Lee, J.S., Koser, A., Rench, D.W., Yeats, A.L., Buckley, B.B., Awschalom, D.D., Samarth, N.: Appl. Phys. Lett. **97**, 262104 (2010)
20. Zeng, Z., Morgan, T.A., Fan, D., Li, C., Hirono, Y., Hu, X., Zhao, Y., Lee, J.S., Wang, J., Wang, Z.M., Yu, S., Hawkridge, M.E., Benamara, M., Salamo, G.J.: AIP Adv. **3**, 072112 (2013)
21. He, L., Kou, X., Wang, K.L.: Phys. Status Solidi (RRL). **7**, 1–2 (2013)
22. Seixas, L., West, D., Fazzio, A., Zhang, S.B.: Nat. Commun. **6**, 7630 (2015)
23. Huang, Y.Q., Song, Y.X., Wang, S.M., Buyanova, I.A., Chen, W.M.: Nat Commun. **8**, 15401 (2017)
24. Hosur, P.: Phys. Rev. **B83**, 035309 (2011)
25. Qu, D.: Spin-Based Broadband Terahertz Radiation from Topological Insulators. Lawrence Livermore National Laboratory, LDRD Annual Report (2017)
26. Lee, K.-K., Myers, T.H.: J. Vasc. Sci. Technol. A. **33**, 031602 (2015)
27. Tsipas, P., Xenogiannopoulou, E., Kassavetis, S., Tsoutsou, D., Golias, E., Bazioti, C., Dimitrakopulos, G.P., Komninou, P., Liang, H., Caymax, M., Dimoulas, A.: ACS Nano. **8**, 6614 (2014)
28. Xenogiannopoulou, E., Tsipas, P., Aretouli, K.E., Tsoutsou, D., Giamini, S.A., Bazioti, C., Dimitrakopulos, G.P., Komninou, P., Brems, S., Huyghebaert, C., Raduc, I.P., Dimoulas, A.: Nanoscale. **7**, 7896 (2015)
29. Chaturvedi, P., Chouksey, S., Banerjee, D., Ganguly, S., Saha, D.: Appl. Phys. Lett. **107**, 192105 (2015)
30. Pang, M.Y., Li, W.S., Wong, K.H., Surya, C.: J. Non-Cryst. Solids. **354**, 4238 (2008)
31. Zhang, X., Wang, J., Zhang, S.-C.: Phys. Rev. **B82**, 245107 (2010)
32. Luo, L., Zheng, K., Zhang, T., Liu, Y.H., Yu, Y., Lu, R., Qiu, H., Li, Z., Huang, J.C.A.: J. Mater. Chem. C. **3**, 9154 (2015)
33. Huang, S.-M., Huang, S.-J., Yan, Y.-J., Yu, S.-H., Chou, M., Yang, H.-W., Chang, Y.-S., Chen, R.-S.: Sci. Rep. **7**, 45413 (2017)
34. Sharma, A., Bhattacharyya, B., Srivastava, A.K., Senguttuvan, T.D., Husale, S.: Sci. Rep. **6**, 19138 (2015)
35. Liu, J., Li, Y., Song, Y., Ma, Y., Chen, Q., Zhu, Z., Lu, P., Wang, S.: Appl. Phys. Lett. **110**, 141109 (2017)
36. Yao, J., Shao, J., Wang, Y., Zhao, Z., Yang, G.: Nanoscale 2015. **7**, 12535 (2015)
37. Qiao, H., Yuan, J., Xu, Z., Chen, C., Lin, S., Wang, Y., Song, J., Liu, Y., Khan, Q., Hoh, H.Y., Pan, C.-X., Li, S., Bao, Q.: ACS Nano. **9**, 1886 (2015)
38. Yao, J., Zheng, Z., Yang, G.: J. Mater. Chem. **C4**, 7831 (2016)
39. Tian, W., Yu, W., Shi, J., Wang, Y.: Materials. **10**, 814 (2017)
40. Yue, Z., Cai, B., Wang, L., Wang, X., Gu, M.: Sci. Adv. **2**, e1501536 (2016)
41. Wei, P., Wang, Z., Liu, X., Aji, V., Shi, J.: Phys. Rev. B. **85**, 201402(R) (2012)
42. Cho, S., Butch, N.P., Paglione, J., Fuhrer, M.S.: Nano Lett. **12**(1), 469 (2012)
43. Zhu, H., Richter, C.A., Zhao, E., Bonevich, J.E., Kimes, W.A., Jang, H.-J., Yuan, H., Li, H., Arab, A., Kirillov, O., Maslar, J.E., Ioannou, D.E., Li, Q.: Sci. Rep. **3**, 1757 (2013)

44. Inhofer, A., Duffy, J., Boukhicha, M., Bocquillon, E., Palomo, J., Watanabe, K., Taniguchi, T., Estève, I., Berroir, J.M., Fève, G., Plaçais, B., Assaf, B.A.: Phys. Rev. **A9**, 024022 (2018)
45. Banerjee, S.K., Register, L.F., Macdonald, A., Sahu, B.R., Jadaun, P., Chang, J.: Topological insulator-based field-effect transistor. Patent No. 8,629,427, 14 January 2014
46. Fatemi, V., Hunt, B., Steinberg, H., Eltinge, S.L., Mahmood, F., Butch, N.P., Watanabe, K., Taniguchi, T., Gedik, N., Ashoori, R.C., Jarillo-Herrero, P.: Phys. Rev. Lett. **113**, 206801 (2014)
47. Ikeda, S., Hayakawa, J., Lee, Y.M., Matsukura, F., Ohno, Y., Hanyu, T., Ohno, H.: IEEE Trans. Electron. Devices. **54**, 991 (2007)
48. Miron, I.M., Garello, K., Gaudin, G., Zermatten, P.-J., Costache, M.V., Auffret, S., Bandiera, S., Rodmacq, B., Schuhl, A., Gambardella, P.: Nature. **476**, 189 (2011)
49. Wang, K.L., Alzate, J.G., Amiri, P.K.: J. Phys. D Appl. Phys. **46**, 074003 (2013)
50. Ramaswamy, R., Lee, J.M., Cai, K., Yang, H.: Appl. Phys. Lett. **5**, 031107 (2018)
51. Liu, L., Pai, C.-F., Ralph, D.C., Buhrman, R.A.: PRL. **109**, 186602 (2012)
52. Huang, L., He, S., Yap, Q.J., Lim, S.T.: Appl. Phys. Lett. **113**, 022402 (2018)
53. Wang, Y., Deorani, P., Banerjee, K., Koirala, N., Brahlek, M., Oh, S., Yang, H.: Phys. Rev. Lett. **114**, 257202 (2015)
54. Fan, Y., Upadhyaya, P., Kou, X., Lang, M., Takei, S., Wang, Z., Tang, J., He, L., Chang, L.-T., Montazeri, M., Yu, G., Jiang, W., Nie, T., Schwartz, R.N., Tserkovnyak, Y., Wang, K.L.: Nat. Mater. **13**, 699 (2014)
55. Melnik, A.R., Lee, J.S., Richardella, A., Grab, J.L., Mintun, P.J., Fischer, M.H., Vaezi, A., Manchon, A., Kim, E.-A., Samarth, N., Ralph, D.C.: Nature. **511**, 449 (2014)
56. Wang, Y., Zhu, D., Wu, Y., Yang, Y., Yu, J., Ramaswamy, R., Mishra, R., Shi, S., Elyasi, M., Teo, K.-L., Wu, Y., Yang, H.: Nat. Commun. **8**, 1364 (2017)
57. Kiselev, S.I., Sankey, J.C., Krivorotov, I.N., Emley, N.C.: Nature. **425**, 380 (2003)
58. Mancoff, F.B., Rizzo, N.D., Engel, B.N., Tehrani, S.: Nature. **437**, 393 (2005)
59. Slavin, A.N., Tiberkevich, V.S.: Phys. Rev. **B74**, 104401 (2006)
60. Grollier, J., Cros, V., Fert, A.: Phys. Rev. **B73**, 060409 (R) (2006)
61. Elyasi, M., Bhatia, C.S., Yang, H.: J. Appl. Phys. **117**, 063907 (2015)
62. Turtle, J., Buono, P.-L., Palacios, A., Dabrowski, C., In, V., Longhini, P.: Phys. Rev. **B95**, 144412 (2017)
63. Qu, T., Victora, R.H.: Sci. Rep. **5**, 11462 (2015)
64. Georges, B., Grollier, J., Cros, V., Fert, A.: Appl. Phys. Lett. **92**, 232504 (2008)
65. Yokoyama, T.: Phys. Rev. **B84**, 113407 (2011)
66. Duan, X., Li, X.-L., Kim, K.W.: Phys. Rev. **B92**, 115429 (2015)
67. Semenov, Y.G., Duan, X., Kim, K.W.: Phys. Rev. **B89**, 201405(R) (2014)
68. Kiehl, R.A., Ohshima, T.: Appl. Phys. Lett. **67**, 2494 (1995)
69. Ohshima, T., Kiehl, R.A.: J. Appl. Phys. **80**, 912 (1996)
70. Wang, C.-Z., Xu, H.-Y., Rizzo, N.D., Kiehl, R.A., Lai, Y.-C.: Phys. Rev. Appl. **10**, 064003 (2018)
71. Yue, Z., Gu, M.: Frontiers in Optics, Optical Society of America, p. FTu3F.1 (2016)
72. Sun, L., Lin, Z., Peng, J., Weng, J., Huang, Y., Luo, Z.: Sci. Rep. **4**, 4794 (2014)
73. Yue, Z., Wang, X., Gu, M.: Topological insulator materials for advanced optoelectronic devices, arXiv:1802.07841v1 [physics.optics] (2018)

Index

V. Litvinov, *Magnetism in Topological Insulators*,
https://doi.org/10.1007/978-3-030-12053-5

Printed in the United States
By Bookmasters